# Advances in Intelligent Systems and Computing

Volume 865

**Series editor**

Janusz Kacprzyk, Polish Academy of Sciences, Warsaw, Poland
e-mail: kacprzyk@ibspan.waw.pl

The series "Advances in Intelligent Systems and Computing" contains publications on theory, applications, and design methods of Intelligent Systems and Intelligent Computing. Virtually all disciplines such as engineering, natural sciences, computer and information science, ICT, economics, business, e-commerce, environment, healthcare, life science are covered. The list of topics spans all the areas of modern intelligent systems and computing such as: computational intelligence, soft computing including neural networks, fuzzy systems, evolutionary computing and the fusion of these paradigms, social intelligence, ambient intelligence, computational neuroscience, artificial life, virtual worlds and society, cognitive science and systems, Perception and Vision, DNA and immune based systems, self-organizing and adaptive systems, e-Learning and teaching, human-centered and human-centric computing, recommender systems, intelligent control, robotics and mechatronics including human-machine teaming, knowledge-based paradigms, learning paradigms, machine ethics, intelligent data analysis, knowledge management, intelligent agents, intelligent decision making and support, intelligent network security, trust management, interactive entertainment, Web intelligence and multimedia.

The publications within "Advances in Intelligent Systems and Computing" are primarily proceedings of important conferences, symposia and congresses. They cover significant recent developments in the field, both of a foundational and applicable character. An important characteristic feature of the series is the short publication time and world-wide distribution. This permits a rapid and broad dissemination of research results.

More information about this series at http://www.springer.com/series/11156

Jezreel Mejia · Mirna Muñoz
Álvaro Rocha · Adriana Peña
Marco Pérez-Cisneros
Editors

# Trends and Applications in Software Engineering

Proceedings of the 7th International
Conference on Software Process Improvement
(CIMPS 2018)

 Springer

*Editors*
Jezreel Mejia
Unidad Zacatecas
Centro de Investigación en Matemáticas
A. C.
Zacatecas, Zacatecas, Mexico

Mirna Muñoz
Unidad Zacatecas
Centro de Investigación en Matemáticas
A. C.
Zacatecas, Zacatecas, Mexico

Álvaro Rocha
Departamento de Engenharia Informática
Universidade de Coimbra
Coimbra, Portugal

Adriana Peña
Departamento de Ciencias Computacionales
Universidad de Guadalajara
Guadalajara, Jalisco, Mexico

Marco Pérez-Cisneros
Centro Universitario de Ciencias Exactas
Universidad de Guadalajara
Guadalajara, Jalisco, Mexico

ISSN 2194-5357        ISSN 2194-5365   (electronic)
Advances in Intelligent Systems and Computing
ISBN 978-3-030-01170-3        ISBN 978-3-030-01171-0   (eBook)
https://doi.org/10.1007/978-3-030-01171-0

Library of Congress Control Number: 2018955572

This Springer imprint is published by the registered company Springer Nature Switzerland AG
The registered company address is: Gewerbestrasse 11, 6330 Cham, Switzerland

# Introduction

This book contains a selection of papers accepted for presentation and discussion at the 2018 International Conference on Software Process Improvement (CIMPS'18). This conference had the support of the CIMAT A.C. (Mathematics Research Center/Centro de Investigación en Matemáticas), CUCEI (Centro Universitario de Ciencias Exactas e Ingenierías de la Universidad de Guadalajara, Jalisco, México), AISTI (Iberian Association for Information Systems and Technologies/Associação Ibérica de Sistemas e Tecnologas de Informação), ReCIBE (Revista electrónica de Computación, Informática, Biomédica y Electrónica), ROPRIN (Red de Optimización de Procesos Industriales), ORACLE, and IBM. It took place at CUCEI, Guadalajara, Jalisco, México, on October 17–19, 2018.

The International Conference on Software Process Improvement (CIMPS) is a global forum for researchers and practitioners that present and discuss the most recent innovations, trends, results, experiences, and concerns in the several perspectives of software engineering with clear relationship but not limited to software processes, security in information and communication technology, and big data field. One of its main aims is to strengthen the drive toward a holistic symbiosis among academy, society, industry, government, and business community, promoting the creation of networks by disseminating the results of recent research in order to align their needs. CIMPS'18 was built on the successes of CIMPS'12, CIMPS'13, CIMPS'14, which took place on Zacatecas, Zac; CIMPS'15 which took place on Mazatlán, Sinaloa; CIMPS'16 which took place on Aguascalientes, Aguascalientes, México; and the last edition CIMPS'17 which took place again on Zacatecas, Zac, México.

The program committee of CIMPS'18 was composed of a multidisciplinary group of experts and those who are intimately concerned with software engineering and information systems and technologies. They have had the responsibility for evaluating, in a 'blind review' process, the papers received for each of the main themes proposed for the conference: organizational models, standards and methodologies; knowledge management; software systems, applications and tools; information and communication technologies and processes in non-software domains (mining, automotive, aerospace, business, health care, manufacturing, etc.) with a demonstrated relationship to software engineering challenges.

CIMPS'18 received contributions from several countries around the world. The papers accepted for presentation and discussion at the conference are published by Springer (this book), and extended versions of best selected papers will be published in relevant journals, including SCI/SSCI and Scopus-indexed journals.

We acknowledge all those who contributed to the staging of CIMPS'18 (authors, committees, and sponsors); their involvement and support are very much appreciated.

October 2018

Jezreel Mejia
Mirna Muñoz
Álvaro Rocha
Adriana Peña
Marco Pérez-Cisneros

# Organization

## Conference

### General Chairs

Jezreel Mejía                      Mathematics Research Center, Research
                                   Unit Zacatecas, Mexico
Mirna Muñoz                        Mathematics Research Center, Research
                                   Unit Zacatecas, Mexico

The general chairs and co-chair are researchers in Computer Science at the Research Center in Mathematics, Zacatecas, México. Their research field is software engineering, which focuses on process improvement, multi-model environment, project management, acquisition and outsourcing process, solicitation and supplier agreement development, agile methodologies, metrics, validation and verification, and information technology security. They have published several technical papers on acquisition process improvement, project management, TSPi, CMMI, multi-model environment. They have been members of the team that has translated CMMI-DEV v1.2 and v1.3 to Spanish.

### General Support

CIMPS General Support represents centers, organizations, or networks. These members collaborate with different European, Latin American, and North American organizations. The following people have been members of the CIMPS conference since its foundation for the last 7 years.

Cuauhtémoc Lemus Olalde          Head of CIMAT Unit Zacatecas, Mexico
Gonzalo Cuevas Agustín           Polytechnical University of Madrid, Spain
Jose A. Calvo-Manzano Villalón   Polytechnical University of Madrid, Spain

Tomas San Feliu Gilabert                    Polytechnical University of Madrid, Spain
Álvaro Rocha                                Universidade de Coimbra, Portugal

## Local Committee

CIMPS established a local committee from the University of Guadalajara, Jalisco,
Mexico, and the Mathematics Research Center; Research Unit Zacatecas, Mexico.
The list below comprises the local committee members.

| | |
|---|---|
| Adriana Peña Pérez Negrón (Local Chair) | CUCEI |
| Graciela Lara López (Local Co-chair) | CUCEI |
| María Elena Romero Gastelu (Public Relations) | CUCEI |
| María Patricia Ventura Núñez (Local Staff) | CUCEI |
| Carlos Alberto López Franc (Local Staff) | CUCEI |
| Elsa Estrada Guzmán (Local Staff) | CUCEI |
| Marco Antonio Pérez Cisneros (Academic Support) | CUCEI |
| Himer Ávila George (Local Staff) | CUVALLES |
| José Antonio Cervantes Álvarez (Local Staff) | CUVALLES |
| Sonia López Ruiz (Local Staff) | CUVALLES |
| Miguel Ángel De la Torre Gómora (Local Staff) | CUVALLES |
| Isaac Rodríguez Maldonado (Support) | CIMAT Unit Zacatecas, Mexico |
| Saúl Ibarra Luevano (Support) | CIMAT Unit Zacatecas, Mexico |
| Ana Patricia Montoya Méndez (Support) | CIMAT Unit Zacatecas, Mexico |
| Héctor Octavio Girón Bobadilla (Support) | CIMAT Unit Zacatecas, Mexico |

## Scientific Program Committee

CIMPS established an international committee of selected well-known experts in
software engineering who are willing to be mentioned in the program and to review
a set of papers each year. The list below comprises the scientific program committee
members.

| | |
|---|---|
| Adriana Peña Pérez-Negrón | University of Guadalajara, Mexico |
| Alejandro Calderón | University of Cádiz, Spain |
| Alejandro Rodríguez González | Polytechnical University of Madrid, Spain |
| Alejandra García Hernández | Autonomous University of Zacatecas, Mexico |
| Álvaro Rocha | Universidade de Coimbra, Portugal |
| Ángel M. García Pedrero | Polytechnical University of Madrid, Spain |
| Antoni Lluis Mesquida Calafat | University of Islas Baleares, Spain |
| Antonio de Amescua Seco | University Carlos III of Madrid, Spain |
| Benjamín Ojeda Magaña | University of Guadalajara, Mexico |
| Carla Pacheco | Technological University of Mixteca, Oaxaca, Mexico |
| Carlos Abraham Carballo Monsivais | CIMAT Unit Zacatecas, Mexico |
| Carlos Lara Álvarez | CIMAT Unit Zacatecas, Mexico |
| Edgar Oswaldo Díaz | INEGI, Mexico |
| Edrisi Muñoz Mata | CIMAT Unit Zacatecas, Mexico |
| Elisabet Cápon | Swiss Federal Institute of Technology, Zürich (ETHZ), Switzerland |
| Eleazar Aguirre Anaya | National Polytechnical Institute, Mexico |
| Fernando Moreira | University of Portucalense, Portugal |
| Gabriel A. García Mireles | University of Sonora, Mexico |
| Giner Alor Hernández | Technological University of Orizaba, Mexico |
| Gloria P. Gasca Hurtado | University of Medellin, Colombia |
| Graciela Lara López | University of Guadalajara, Mexico |
| Gonzalo Cuevas Agustín | Polytechnical University of Madrid, Spain |
| Gonzalo Luzardo | Higher Polytechnic School of Litoral, Ecuador |
| Gustavo Illescas | National University of Central Buenos Aires Province, Argentina |
| Himer Ávila George | University of Guadalajara, Mexico |
| Hugo Arnoldo Mitre | CIMAT Unit Zacatecas, Mexico |
| Hugo O. Alejandrez-Sánchez | National Center for Research and Technological Development, CENIDET, Mexico |
| Iván García Pacheco | Technological University of Mixteca, Oaxaca, Mexico |
| Jezreel Mejía Miranda | CIMAT Unit Zacatecas, Mexico |
| José Alberto Benítez Andrades | University of Lion, Spain |
| Jose A. Calvo-Manzano Villalón | Polytechnical University of Madrid, Spain |
| José Antonio Cervantes Álvarez | University of Guadalajara, Mexico |
| José Luis Sánchez Cervantes | Technological University of Orizaba, Mexico |

# Contents

# Organizational Models, Standards and Methodologies

# Extending ISO/IEC 29110
# with Sustainability Tasks

Gabriel Alberto García-Mireles[1]([⊠])
and Miguel Ehécatl Morales-Trujillo[2]

[1] Departamento de Matemáticas,
Universidad de Sonora, Hermosillo, Sonora, Mexico
mireles@mat.uson.mx
[2] Computer Science and Software Engineering Department,
University of Canterbury, Christchurch, New Zealand
miguel.morales@canterbury.ac.nz

**Abstract.** Sustainability is an aspect to be considered by software development organizations because it is believed that software systems are means to support sustainable development. Currently, few processes, methods and tools exist to practice sustainability design within software engineering, and they are scattered across different application domains and life cycle stages. With the purpose of providing a set of practices in order to address sustainability goals during software development, this paper proposes a set of sustainability tasks to be implemented by very small software organizations. The tasks were derived from literature containing sustainability practices. The latter were organized as tasks and integrated into the ISO/IEC 29110 Basic Profile processes: project management and software implementation. The proposed sustainability tasks were initially validated against two previously reported studies on environmentally sustainable software development.

**Keywords:** Environmental sustainability · Green software
Software process improvement · Sustainability tasks · ISO/IEC 29110

## 1  Introduction

Within software engineering (SE) community, sustainability is a topic of recent interest as literature shows work in diverse domains and throughout software development life cycle stages [1]. In order to address sustainability in software projects, researchers have suggested that software developers should consider its dimensions (economic, social, environment, at least) as well as the direct and indirect impacts on sustainable development [2, 3].

Sustainable software is defined as follows: "software, whose impacts on economy, society, human being, and environment that result from development, deployment and usage of the software are minimal and/or which has a positive effect on sustainable development" [4]. Although the SE sustainability research community suggests focusing on all the sustainability dimensions [5], the majority of research work has been conducted in the environmental dimension [6]. Environmental issues can also be

© Springer Nature Switzerland AG 2019
J. Mejia et al. (Eds.): CIMPS 2018, AISC 865, pp. 3–13, 2019.
https://doi.org/10.1007/978-3-030-01171-0_1

treated as the green dimension of sustainable development [7]. In this dimension, the strategies to follow are minimizing the use of resources as well as reducing energy consumption.

Research in sustainability from a software engineering perspective is in its initial stage regarding both its definition and related software development practices. On the first issue, sustainability is treated as a quality goal [8], but it is difficult to identify its components in order to establish appropriate indicators and measures [9]. On the issue of sustainability practices, Chitchyan et al. [10] pointed out that software practitioners need methodologies to carry out sustainable design within SE. Similarly, Manotas et al. [11] mentioned that practitioners lack the necessary information and support infrastructure to develop green software. In addition, there are multiple factors that influence the extent to which the sustainability can be achieved in a software product [12]. Indeed, sustainability is a systemic aspect that has multiple dimensions and requires actions on multiple levels [5].

Considering the need of developing a systemic approach to address sustainability during a software development life cycle, this paper suggests a set of sustainability tasks, i.e., tasks aiming to achieve sustainability goals, based on sustainability practices discussed in surveys, literature reviews, and papers addressing sustainability practices with a software process focus. The sustainability tasks are presented as an extension of the ISO/IEC 29110 Basic Profile process activities [13]. A preliminary validation was carried out through an attempt to categorize practices from studies on green software development by means of our proposal.

The paper is structured as follows. Section 2 describes the relevant literature about sustainability and software processes. Section 3 briefly presents the methodology followed in this research. Section 4 describes our proposal focusing on sustainability tasks integrated into the ISO/IEC 29110 processes, while Sect. 5 presents an initial validation. Discussion of results is described in Sect. 6. Finally, conclusions and further work are addressed in Sect. 7.

## 2   Related Work

This section provides a sustainability in SE background and briefly presents ISO/IEC 29110 to support the proposal presented herein. In this paper, the term 'practices' refers to actions carried out by practitioners or those reported by case studies. 'Activity' and 'task' terms are used as defined in the ISO/IEC 29110 [13]. The former term refers to a set of tasks while the latter describes a recommended action intended to accomplish a process objective.

### 2.1   Background

SE and sustainability literature reviews reflect trends in methods and practices. In the period from 2006 to 2012, few papers addressed this topic and researchers founding little methodological guidance to support sustainability [14]. In a follow up study, researchers found sustainability studies in the areas of software process, software design and software quality [1]. However, the authors concluded that there is little evidence of sustainability practices validation in industrial settings [1].

In another literature review about software process and environmental sustainability [15], the author reported few theoretical papers. In general, the selected papers focused on identifying software life cycle stages that should address sustainability aspects [16, 17] and on developing sustainability processes based on the ISO/IEC 12207 [18].

With regard to software life cycle proposals, Naumann et al. [16] described the GreenSoft Model whose goals are oriented towards achieving better sustainable development through software. This model considers both direct and indirect impacts on environmental, social, and economic dimensions. The product life cycle consists of development, usage, and end of life stages; however, no detailed structure of each of them is provided. Also, the model includes an example of a software process that fits the development stage. The description is carried out through: sustainability reviews, process assessment, a sustainability journal (containing a log of improvement effects) and a sustainability retrospective [16].

Taking into account the characteristics of the GreenSoft Model, Mahmoud and Ahmad [17] proposed a software life cycle model, which involves requirements and testing stages, for addressing green computing. In addition, each life cycle stage consists of actions and recommendations. On the other hand, Lami et al. [18] defined three processes based on the ISO/IEC 12207 [19]. They are: a sustainability management process, a sustainability engineering process and a sustainability qualification process. However, these processes are discussed at a high level of abstraction in terms of process purposes and outcomes.

A methodological support for addressing sustainability is scattered across software development life cycle stages and until now, there are few validated proposals in industrial settings [1, 20]. Indeed, sustainability in SE is still in an immature stage [6]. Thus, in order to organize practices addressing sustainability, we decided to present them as an extension to ISO/IEC 29110 [13]. The sustainability should be addressed by all software organizations and considered in all software projects [5]. Therefore, the proposal presented herein contributes to addressing sustainability within very small organizations. In addition, given the immature state of sustainability in SE, we believe that the ISO/IEC 29110 Basic Profile [13] is an appropriate framework to organize current sustainability practices.

## 2.2  ISO/IEC 29110 Basic Profile

The ISO/IEC 29110 Basic Profile [13] consists of two processes: project management and software implementation, which are derived from the ISO/IEC 12207 software life cycle process [19] taking into account very limited resources of software organizations composed of up to 25 persons. The expected benefits of implementing ISO/IEC 29110 are an increase of customer satisfaction and product quality together with a minimization of development costs.

The purpose of the project management process is to establish a plan to support software development activities in order to fulfill expected project goals. It is consists of the following activities: planning, executing, assessment and control, and closure. The purpose of the software implementation process is to systematically develop a software product through the activities of software implementation initiation, software requirements analysis, software architectural and detailed design, software construction, software integration and tests, and product delivery. While the project management process consists of four activities and 26 tasks, the software implementation process comprises six activities and 41 tasks.

## 3   Approach to Define a Sustainability Extension

In order to develop this proposal, the following steps were carried out:

1. Identifying relevant papers addressing sustainability related practices. In order to provide a full view of the software process and to identify practices instead of techniques, the papers can be gathered from systematic literature reviews, surveys conducted in industrial settings as well as papers that address sustainability practices for more than one software life cycle stage. The selected papers are presented in Sect. 4.
2. Mapping identified sustainability practices to ISO/IEC 29110. A template that considers both the ISO/IEC 29110 [13] activities and tasks was developed to guide the mapping process. Each selected paper was reviewed and text fragments addressing a practice or recommendation were extracted. The template with process activities was filled in with the related text fragments. The mapping was carried out by the first author and verified by the second author. Based on the identified sustainability practices, *sustainability tasks* were proposed (Tables 1 and 2).
3. Defining a sustainability related role and products. A sustainability expert role is needed to support sustainability related practices. Their responsibilities are described in Sect. 4. The ISO/IEC 29110 [13] products were reviewed and an enhanced proposal is presented to include sustainability aspects (Table 3).
4. Carrying out an initial validation. The initial validation is focused on assessing to what extent the *sustainability tasks* are useful for categorizing sustainability practices used in case studies or experience-based suggestions. Two papers addressing sustainability practices during software development were selected. The first paper provides a framework of recommended practices for developing green software, mainly targeted for software design and construction stages [21]. The second paper describes a case study which addresses practices for eliciting sustainability requirements [22]. The mapping procedure was similar to the described in bullet 2, where text fragments extracted from these two papers were labeled with corresponding activities from ISO/IEC 29110. The second author verified the identification and classification of text fragments. Issues were resolved through a virtual peer debriefing.

**Table 1.** Sustainability tasks for project management process

| Activity | Sustainability task | References |
|---|---|---|
| Project planning | 1. Review business and software product goals considering sustainability<br>2. Determine sustainability scope in terms of dimensions, stakeholders, system boundaries, and time span<br>3. Identify and document sustainability risks<br>5. Define sustainability roles, responsibilities and specific tasks<br>6. Identify sustainability training needs<br>7. Review sustainability goals with customer<br>8. Establish a sustainability repository | [10–12, 17, 23, 24] |
| Project plan execution | 9. Analyze change requests for sustainability goals<br>10. Review with development team sustainability goals' issues and mitigation actions<br>11. Review with customer and other stakeholders changes on sustainability goals | [13, 20] |
| Project assessment and control | 12. Evaluate project progress considering sustainability goals<br>13. Establish corrective actions when sustainability risks emerge | [17, 24] |
| Project closure | 14. Inform customer on the extent sustainability goals were addressed in software product | [13, 24] |

**Table 2.** Sustainability tasks for software implementation process

| Activity | Sustainability task | Reference |
|---|---|---|
| Software implementation initiation | 1. Review with work team sustainability goals in current project<br>2. Set up the implementation environment considering means to support sustainability | [10, 11, 13, 23] |
| Software requirements analysis | 3. Identify and document sustainability requirements considering sustainability goals<br>4. Analyze sustainability requirements considering their feasibility, risk, and potential interdependencies among dimensions | [11, 17, 20, 23] |

*(continued)*

**Table 2.** (*continued*)

| Activity | Sustainability task | Reference |
|---|---|---|
|  | 5. Review that sustainability requirements are consistent with product description and are testable |  |
| Software architectural and detailed design | 6. Based on sustainability dimensions addressed, determine sustainability criteria to define the software architecture<br>7. Document software architecture based on sustainability criteria<br>8. Verify that software design meets sustainability criteria and requirements<br>9. Develop test cases for sustainability goals | [11, 17, 20, 23, 24] |
| Software construction | 10. Construct software components considering sustainability criteria<br>11. Test software components considering sustainability requirements | [11, 17, 20] |
| Software integration and tests | 12. Execute sustainability test cases<br>13. Determine to what extent the system achieves sustainability goals (e.g.: develop energy profiles) | [11, 17] |
| Product delivery | 14. Document, for maintenance activities, tools and procedures used to achieve sustainability goals | [17, 24] |

**Table 3.** Sustainability items to be added to existing work products

| Product | Sustainability related action |
|---|---|
| 4. Maintenance documentation | Add a section to describe the environment (compilers configuration, static analysis tools, energy profiler tools, among others) for defining sustainability requirements, energy profiles and testing |
| 6. Product operation guide | The product describes criteria for operational use and it could include sustainability scenarios and recommended operation modes that contribute to energy efficiency or other sustainability goals |
| 7. Progress status record | Include a record of actual results about sustainability goals against planned goals |
| 8. Project plan | Include sustainability objectives of the project. Deliverables and tasks with which sustainability goals are addressed should be defined. Resources to address sustainability goals as well as the required training should be described. Identify a sustainability related repository. Identify project risks considering sustainability in their both direct and indirect impacts as well as in the sustainability dimensions |
| 11. Requirements specification | Add a section to describe sustainability requirements considering their impacts and dimensions. Include potential interaction among sustainability requirements and nonfunctional requirements |

(*continued*)

**Table 3.**  (*continued*)

| Product | Sustainability related action |
|---|---|
| 15. Software design | Include a description of the extent sustainability goals and requirements to guide software architecture definition. Within software components, describe sustainability aspects they support |
| 16. Software user documentation | In the operational environment section, it should include a description of the way sustainability goals can be achieved with current software version. It could include sustainability risks, warnings and notes |
| 17. Statement of work | It should include sustainability objectives of the project |
| 18. Test cases and test procedures | Include in test cases and procedures the means to identify sustainability related issues. They could consider energy profiles |

# 4   Sustainability Tasks Integrated into ISO/IEC 29110

The articles considered to identify sustainability practices can be classified into surveys conducted among practitioners [10, 11, 23], proposals focused on software processes [17, 24] and models derived from literature reviews [12, 20]. They are referenced in Tables 1 and 2 grouped by process activities from ISO/IEC 29110 [13].

In the surveys group, Groher and Weinreich [23] presented a qualitative interview study, targeted to software project team leaders, with the aim of understanding how sustainability is currently managed in software development projects. The authors identified influencing factors, problems and measures they took to improve sustainability. In another qualitative study, Manotas et al. [11] conducted a survey targeted to 464 practitioners, with the purpose of identifying current practices used by software engineers to address green SE. They provided a suggestion to develop suitable software. On the other hand, Chitchyan et al. [10] explored the perceptions and attitudes of requirement engineers towards sustainability. They identified concerns practitioners faced and mitigation strategies based on sustainability design principles.

In the software process related proposals, Dick et al. [24] proposed an agile extension to software development process to integrate sustainability practices both into software process and software product. The proposed practices are complemented by some tools and guidelines. Mahmoud and Ahmad [17] proposed sustainability practices that could be addressed at requirements, design, coding and testing stages. In addition, the authors proposed a green analysis stage to determine the greenness of each increment of the system under development [17].

Based on literature reviews, Chitchyan et al. [12] analyzed practices in software product line domain to understand sustainability concerns software developers deal with. The authors grouped the factors that influence sustainability according to dimensions and analyzed both the cross-dimensional dependencies and the influence of stakeholders [12]. Finally, a mapping study about interactions between software quality and environmental sustainability [20] organizes methods found in SE knowledge areas.

In addition to the roles specified in ISO/IEC 29110 [13], this proposal suggests that a sustainability expert role is needed. This role has gathered knowledge about sustainability in SE and they have the abilities to apply it in software projects. In addition, he or she is responsible for supporting the development team and project manager as well as for keeping up the sustainability repository.

Table 1 depicts the sustainability tasks for the project management process. The first column displays an activity from ISO/IEC 29110 while the second column shows proposed sustainability tasks. The third column presents references. For instance, the 'Task 6. Identify sustainability training needs' was derived from references such as [23] who established that "training of personnel was regarded as very important. Training multiple persons on the topics decreases the key person risk". Similarly, [10] suggested that "it is necessary to educate … practitioners on the subject of sustainability design through formal education …, practice guidelines, demonstrative examples/case studies, and alike."

Table 2 presents proposed sustainability tasks for software implementation process. Task 4 "Analyze sustainability requirements considering their feasibility, risk, and potential interdependencies among dimensions" is based on Mahmoud and Ahmad model [17] who recommended performing risk analysis by taking into account energy efficiency and suggesting that feasibility analysis can be of help in identifying benefits related to improving energy efficiency. To support different sustainability goals, García-Mireles et al. [20] reported the usage of modeling languages to identify and analyze sustainability goals as well as the prioritization of sustainability requirements when conflicts among sustainability goals emerge.

In this proposal, we use the same ISO/IEC 29110 [13] products to address sustainability. Table 3 describes the products that need to include sustainability goals and specifications. The product number corresponds to the number described in ISO/IEC 29110 [13]. Thus, the ISO/IEC 29110 task list tables can be used to relate products with proposed sustainability tasks.

## 5   Initial Validation

To validate the feasibility of the proposed set of tasks, two papers (presented in Sect. 3, bullet 4) were reviewed in order to identify recommended practices to address sustainability and map them to proposed sustainability tasks. In the first paper [21] we identified 30 text segments that were categorized among activities such as software architecture and detailed design, software construction, among others (Table 4, Case A). The recommended practices were categorized in eight sustainability tasks from

**Table 4.** Software implementation process activities addressed by text fragments

| Case study | Activity | Number of text fragments |
|---|---|---|
| Case A | Software architecture and detailed design | 17 |
| Case A | Software construction | 8 |
| Case A | Software implementation initiation | 3 |
| Case A | Software integration and test | 2 |
| Case B | Software requirements analysis | 20 |

Table 2 (tasks number 2, 6, 7, 8, 9, 10, 11, and 13). The tasks with more text fragments were Task 6 (7 text fragments) and Task 10 (6 text fragments). The former task is centered on defining the software architecture whereas the latter refers to constructing software. An example of a text fragment for task 10 is as follows [21]: "Clean up useless code and data. ...writing to never-read variables and other useless routines (such as repeated conditionals) might consume power purposelessly."

In the second paper [22] we identified 20 text segments that belong to software requirements analysis activity (Table 4, Case B). The text segments were categorized in Task 3 (15 text segments) and Task 4 (5 text segments). Task 3 is related to the identification and documentation of sustainability requirements considering sustainability goals. Task 4 focuses on analyzing sustainability requirements with regard to their feasibility and risk, and potential interdependencies among dimensions. An example of a Task 3 text fragment is as follows [22]: "Online questionnaire...Our goal was to elicit quantitative information about important aspects of sustainability requirements from a sample of involved people which is larger than the number of meal planners interviewed."

## 6  Discussion

This study proposed a sustainability extension to the ISO/IEC 29110 Basic Profile [13]. The sustainability extension addresses 14 tasks to be included in the project management process and the same number of tasks for addressing sustainability in the software implementation process. In comparison with the ISO/IEC 29110 Basic Profile which consists of 67 tasks, the 28 proposed tasks represent around 42% of new tasks. However, our proposal gives visibility to sustainability aspects during software development.

The theoretical proposal was validated by identifying sustainability practices in an existing case study and a framework for developing green software. The 50 text fragments were successfully categorized within 10 of the 14 sustainability tasks of the software implementation process. However, we have not found any reference to sustainability tasks related to the software project management process. In the literature, it is difficult to find empirical evidence for sustainability in software project management [1, 20]. Thus, we need to conduct empirical studies to validate the task extension proposed herein.

Considering internal validity threats, the definition of sustainability tasks was derived from at least two literature sources addressing similar practices. The proposed tasks are described on an abstraction level appropriate to be included into ISO/IEC 29110 [13]. Furthermore, to mitigate the misinterpretation of the proposed sustainability tasks, the second author classified text fragments. As a result, we achieved 86% of classification agreement, and all the inconsistencies were resolved.

# 7  Conclusions

Based on literature sources, this paper proposes a set of 28 sustainability tasks, which were mapped to the ISO/IEC 29110 Basic Profile process activities. An initial validation showed that our proposal can be used to map sustainability practices. However, this work is in an initial stage.

As future work, it is necessary to validate the sustainability tasks with research experts in sustainability and software processes. Afterwards, the sustainability extension should be applied in case studies to assess its practical feasibility and the required effort to implement the proposed tasks.

# References

1. Penzenstadler, B., Raturi, A., Richardson, D., Calero, C., Femmer, H., Franch, X.: Systematic mapping Study on Software Engineering for Sustainability (SE4S). In: ACM International Conference Proceeding Series, pp. 1–10 (2014)
2. Becker, C., Betz, S., Chitchyan, R., Duboc, L., Easterbrook, S.M., Penzenstadler, B., Seyff, N., Venters, C.C.: Requirements: The key to sustainability. IEEE Softw. **33**(1), 56–65 (2016)
3. Penzenstadler, B., Femmer, H.: A generic model for sustainability with process- and product-specific instances. In: GIBSE 2013 - Proceedings of the 2013 Workshop on Green in Software Engineering, Green by Software Engineering, pp. 3–7 (2013)
4. Dick, M., Naumann, S.: Enhancing software engineering processes towards sustainable software product design. In: EnviroInfo, pp. 706–715 (2010)
5. Becker, C., Chitchyan, R., Duboc, L., Easterbrook, S., Penzenstadler, B., Seyff, N., Venters, C.C.: Sustainability Design and Software: The Karlskrona Manifesto. In: Proceedings - International Conference on Software Engineering, pp. 467–476 (2015)
6. Calero, C., Piattini, M.: Puzzling out software sustainability. Sustain. Comput.: Inform. Syst. **16**, 117–124 (2017)
7. Calero, C., Piattini, M.: Introduction to green in software engineering. In: Calero, C., Piattini, M., (eds.). Green in Software Engineering, pp. 3–27 (2015)
8. Lago, P., Koçak, S.A., Crnkovic, I., Penzenstadler, B.: Framing sustainability as a property of software quality. Commun. ACM **58**(10), 70–78 (2015)
9. Venters, C.C., Jay, C., Lau, L.M.S., Griffiths, M.K., Holmes, V., Ward, R.R., Austin, J., Dibsdale, C.E., Xu, J.: Software sustainability: the modern tower of babel. In: CEUR Workshop Proceedings, pp. 7–12. (2014)
10. Chitchyan, R., Becker, C., Betz, S., Duboc, L., Penzenstadler, B., Seyff, N., Venters, C.C.: Sustainability design in requirements engineering: state of practice. In: Proceedings of the 38th International Conference on Software Engineering Companion, pp. 533–542. ACM (2016)
11. Manotas, I., Bird, C., Zhang, R., Shepherd, D., Jaspan, C., Sadowski, C., Pollock, L., Clause, J.: An empirical study of practitioners' perspectives on green software engineering. In: Proceedings of the 38th International Conference on Software Engineering, pp. 237–248. ACM, Austin (2016)
12. Chitchyan, R., Groher, I., Noppen, J.: Uncovering sustainability concerns in software product lines. J. Softw.: Evol. Process. **29**(2), 1–20 (2017)

13. ISO: ISO/IEC TR 29110-5-1-2 Software engineering - Lifecycle profiles for Very Small Entities (VSEs) Part 5-1-2: Management and engineering guide: Generic profile group: Basic profile, Geneva (2011)
14. Penzenstadler, B., Bauer, V., Calero, C., Franch, X.: Sustainability in software engineering: a systematic literature review. In: IET Seminar Digest, pp. 32–41 (2012)
15. García-Mireles, G.A.: Environmental Sustainability in Software Process Improvement: a Systematic Mapping Study. In: International Conference on Software Process Improvement, pp. 69–78. Springer (2016)
16. Naumann, S., Dick, M., Kern, E., Johann, T.: The GREENSOFT Model: a reference model for green and sustainable software and its engineering. Sustain. Comput.: Inform. Syst. 1(4), 294–304 (2011)
17. Mahmoud, S.S., Ahmad, I.: A green model for sustainable software engineering. Int. J. Softw. Eng. Its Appl. 7(4), 55–74 (2013)
18. Lami, G., Fabbrini, F., Fusani, M.: Software sustainability from a process-centric perspective. In: Winkler, D., O'connor, R.V., Messnarz, R. (eds.) Systems, software and services process improvement. European Conference on Software Process Improvement. 301 CCIS, pp. 97–108. (2012)
19. ISO/IEC: Systems and software engineering – software life cycle processes - Redline. ISO/IEC 12207:2008(E) IEEE Std 12207–2008 – Redline, pp. 1–195 (2008)
20. García-Mireles, G.A., Moraga, M.Á., García, F., Calero, C., Piattini, M.: Interactions between environmental sustainability goals and software product quality: a mapping study. Inf. Softw. Technol. 95, 108–129 (2018)
21. Ardito, L., Procaccianti, G., Torchiano, M., Vetrò, A.: Understanding green software development: a conceptual framework. IT Prof. 17(1), 44–50 (2015)
22. Huber, M.Z., Hilty, L.M., Glinz, M.: Uncovering sustainability requirements: an exploratory case study in canteens. In: CEUR Workshop Proceedings, pp. 35–44 (2015)
23. Groher, I., Weinreich, R.: An interview study on sustainability concerns in software development projects. In: 2017 43rd Euromicro Conference on Software Engineering and Advanced Applications (SEAA), pp. 350–358. IEEE (2017)
24. Dick, M., Drangmeister, J., Kern, E., Naumann, S.: Green software engineering with agile methods. In: 2013 2nd International Workshop on Green and Sustainable Software (GREENS), pp. 78–85. IEEE (2013)

# Methodologies, Methods, Techniques and Tools Used on SLR Elaboration: A Mapping Study

Marco Palomino, Abraham Dávila$^{(\boxtimes)}$, and Karin Melendez

Departamento de Ingeniería, Pontificia Universidad Católica del Perú, Lima, Peru
{palomino.marco,abraham.davila,kmelendez}@pucp.edu.pe

**Abstract.** The aim of this study is to perform a Systematic Literature Mapping (SLM) about methodologies, methods, techniques and tools used on the development of Systematic Literature Reviews (SLR). As a result, on the mapping, we expected to find and classify methodologies, methods, techniques and tools commonly used on SLR. In addition, we have considered other contexts such as, Medicine or Education with the purpose of getting multiple methodologies, methods, techniques and tools that allow performing SLR on efficient ways. It is mainly expected to identify techniques related to research questions formulation and the methods used for building search strings in order to get the higher number of studies associated to the research topic. In our study, we found multiple methodologies, methods, techniques and tools already implemented for performing SLRs On the mapping we describe some of them to highlight the most used and referenced studies.

**Keywords:** Systematic Literature Review · Methodologies · Methods
Techniques

## 1 Introduction

The research, as well as other activities, should be supported by methodologies, methods, techniques and tools among others to obtain the expected results in a systematic way. In particular, the Systematic Literature Review (SLR) in Software Engineering domain is a case where research teams use methodologies, techniques and tools to help them on reducing time and effort in the research [1, 2]. The SLR is a type of secondary study [3] based on (i) the selection of primary studies of digital libraries and repositories; (ii) the analysis of the data or evidence collected and (iii) the preparation of the answers to the research questions.

Multiple studies have researched tools or search-strategies on a specific SLR activity [4–6]. In fact, the study [7] is a mapping about tools used in SLRs.

In our research, we expanded the previous studies finding also methodologies, tools, methods and techniques commonly used on SLR process. In addition, this approach will help future researchers to work on SLR efficiently due to it will focus on such activities that are manual and intensive such as Planning and Selection.

© Springer Nature Switzerland AG 2019
J. Mejia et al. (Eds.): CIMPS 2018, AISC 865, pp. 14–30, 2019.
https://doi.org/10.1007/978-3-030-01171-0_2

In this context, the purpose of our study was to identify the methodologies, methods, techniques and tools used in the preparation of an SLR, in different contexts such as medicine or education. Additionally, we wanted to identify the available techniques and tools that improve SLR performance. For this study a Systematic Mapping of the Literature (SML) was carried out in the relevant digital databases such as ACM Digital Library, IEEE Xplore, ProQuest, EBSCO, Emerald and Springer.

The present study is organized as follows: Sect. 2, SLR fundamentals are presented; Sect. 3, describes the methodology and the considerations used in the elaboration of the research questions; Sect. 4, results are presented associated with the research questions; and in Sect. 5, final discussion and future work are presented.

## 2 Systematic Literature Review Fundamentals

Systematic Literature Reviews (SLR) have recently been introduced to the context of Software Engineering [8, 9] as a structured methodology to perform literature reviews based on similar models from other contexts, such as education or medicine [10].

Evidence-Based Software Engineering (EBSE) guides that have been appearing in the academic field allow to systematize a literature review setting out a scientific model, which allows, to identify new areas of research, on deepen existing areas and generate knowledge bases [11, 12] that will help future researches in the Software Engineering area [13, 14].

Systematic Literature studies, such as mappings or revisions, have emerged as a way of synthesizing part of the vast scientific evidence that is increasing as new research in the area that occurs [8]. As it is mentioned on [15], these studies, allowing future researchers to take a point of reference and a conceptual framework for the development of their studies. In addition, the SLRs are useful to identify literature and research gaps relevant in a topic of interest [1, 9]. The three main phases of an SLR include planning, conducting and reporting [11].

## 3 Systematic Literature Mapping

This section presents the SLM fundamentals taken into account and its application in this study. Also, in this section, we describe the planning of the SLM based on guidelines proposed by Petersen described in [16], and recommendation from SLR [8, 11, 17].

### 3.1 Systematic Literature Mapping Fundamentals

A SLM study provides a structure of the research reports and results that have been published using visual summary of their results [17]. It often requires less effort while providing a more solid overview. Previously, studies of systematic mappings in Software Engineering have been recommended mainly in the areas of research where there is a need for relevant primary studies [11, 16].

In addition, in order to frame the research question and define the search string, it was used the PICO (Population, Intervention, Comparison, Outcome) criteria applied to Software Engineering [11]. In this study, we have elaborated eight research questions that are displayed in the Sect. 4. Finally, the Table 1 shows the principal keywords and the Table 2 indicates the search strings elaborated.

**Table 1.** Principal keywords used based on PICO criteria

| | |
|---|---|
| Population | Systematic literature review |
| Intervention | Techniques, practices, tools, methodology |
| Comparison | None |
| Outcome | Evaluation, experiences, classification |

**Table 2.** Search strings

| Data source | Search string |
|---|---|
| IEEE Xplore | ("Systematic Literature" OR "Literature Review") AND ("Methodology" OR "Protocol" OR "Guide") AND ("Conduct*" OR "Perform") |
| ProQuest | ("Systematic Literature" "Literature Review" "search engine") AND (Methodology protocol guide) AND ("conduct" "search") AND (tool*) |
| EBSCO | ("Systematic Literature" "Literature Review") AND (Methodology protocol guide) AND ("conduct" "search") AND (Tool) |
| ACM Digital Library Emerald | ("Systematic Literature" "Literature Review" "search engine") AND (Methodology protocol) AND (conduct search) AND (tools) Abstract:("literature review" AND "perform") AND Anywhere("methodology" AND "tool" AND "search") |
| Springer | "Methodology" AND "Systematic Literature Review" AND ("tools" OR "technique" OR "Conduct" OR "Perform") |

The search strings were adapted to the specific syntax of data sources. Also, asterisks (*) were used in the strings to obtain the greatest number of occurrences without explicitly considering the plurals and words derived from the main concepts. These characters allow the engines of the databases to obtain the different possible combinations based on the base word that the asterisk refers.

## 3.2  SLM Protocol

A SLM protocol was defined to reduce the possibility of researcher bias. This protocol described in [16], which mention several steps that help to conduct a SLM, was structured in five main steps that included:

Step 1:       Studies were selected regarding the execution of search string.

Step 2 and 3:   Both steps were related to the process of including and excluding articles using the criteria defined. Then, we analyzed the titles and abstracts.

Step 4:       Then, the review of article Introduction and Conclusion

Step 5:       At the end of the review, the final articles were verified by peer review to evaluate their exclusion or inclusion in our research

The exclusion and inclusion criteria considered were:

- Inclusion criteria: We have considered academic articles with methodological support (controlled experiments, case studies, and systematic reviews, systematic mapping or others). In addition, we included studies extracted from the mentioned digital databases. Studies will be accepted if in their content propose techniques, guides or methods to conduct an SLR (mainly, the preparation of research questions and search strings). Studies that present cases of success in the use of alternative techniques or protocols were also included.
- Exclusion criteria: We have not considered duplicate studies. Additionally, articles that are contained in subsequent articles were also excluded. Also, we didn't consider studies that are not part of the following types of publications: journals, conferences and digital databases. In addition, studies that mention techniques or procedures that do not present concrete results are omitted. Studies whose title is irrelevant or outside the context of the SLR were excluded. Finally, tertiary studies and conference abstracts weren't considered.

## 3.3  Quality Assessment

Quality assessment of this SLM followed 11 criteria defined by [18] based on [19]. The following are the criteria used in the quality assessment:

- Is this study based on research?
- Is there a clear statement of the aims of the research?
- Is there an adequate description of the context?
- Was the study design appropriate to address the aims of the research?
- Was the selection strategy appropriate to the aims of the research?
- Was there a control group for comparing treatments?
- Was the data collected in a way that addressed the research aims?
- Was the data analysis rigorous enough?

- Has the relationship between researcher and participants been considered as an adequate degree?
- Is there a clear statement of results?
- Is the study relevant for practice or research?

According to [19], these mentioned criteria include three important issues related to quality, which were considered in the quality assessment:

- Rigor: a complete approach was applied to key research methods in the study?
- Credibility: are the results in a meaningful and well-presented way?
- Relevance: how useful are the results to the software industry and the scientific community?

For the assessment, each one of the primary studies obtained after inclusion and exclusion criteria of the SLM protocol was analyzed using the 11 questions defined. The scale used in the assessment had two values ("yes" or "no"). When the answer was affirmative, the criteria had a value of "1"; otherwise, the value was "0". As a result, the minimum result could be "0" and "11" as maximum value.

### 3.4    Data Extraction and Data Synthesis Strategies

The Petersen Guides [16] suggest the exploration of some papers sections in case the abstract is not well-structured or vague. In our case, all primary studies were fully read to answer research question as a part of SML protocol. Then, the primary studies were grouped. Finally, in order to conduct the analysis, a narrative synthesis was defined [20]; especially the "Grouping and Clustering".

### 3.5    Studies Selection

The studies selection process started with the automatic search in July 2017 with the first tests. Then, in August 2017, the last execution was performed. Using a spreadsheet editor, we selected the titles, abstracts and references from studies obtained after executing the search strings in the digital sources. After this step, 2,139 potential studies were identified. In Table 3, we could find the initial results.

**Table 3.** Data sources of the systematic review

| Type | Name of database | Initial results |
|---|---|---|
| Automatic search | IEEE Xplore | 376 |
| | ACM Digital Library | 783 |
| | ProQuest | 143 |
| | EBSCO | 170 |
| | Emerald | 326 |
| | Springer | 341 |

Then, during the selection process, we have removed the duplicated studies initially. Then, after the five steps mentioned in the protocol, 54 studies were identified. In addition, we analyzed the references of the 54 studies in order to get additional studies that could be helpful in the research, this approach is called snowball strategy [21] and it consists on getting more relevant information reviewing the references of initial results [6]. At the end, we added 5 more studies and 59 studies were identified on selection process.

The quality assessment was applied in 59 studies identified in the fourth step and the results are displayed in Appendix B. Then, the studies were reviewed with project members to ensure the suitability of the selected primary studies.

## 4   Results

We identified 59 studies, which are listed in Appendix A. The following section will display the answer of the research questions.

### 4.1   RQ-1. How Is the Evolution of the Number of Publications Related to the Research Topic?

This research question will be answered using two approaches, "by year" and "by country affiliation". Figure 1 shows the studies obtained grouped by the year of publication. The studies were obtained in the 1989–2017 period; additionally, the numbers of publication are increased in the last years, growth in average from 2.86 (2004–2010) to 4.86 (2011–2017). This shows the continuous interest of the research community to improve the SLR performance.

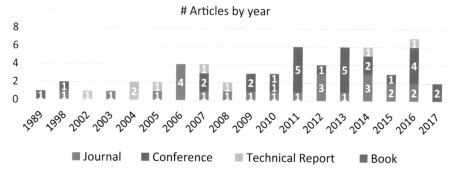

**Fig. 1.**  Number of articles found by year

Regarding the country of affiliation of the article's authors, we display in the Fig. 2 the number of articles obtained by country. As we can see, United Kingdom (UK) and United States of America (USA) are the two countries that provide the most number of studies with almost 50% of the total articles.

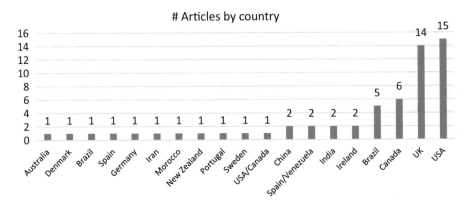

**Fig. 2.** Distribution of articles regarding country affiliation

**Table 4.** Number of articles regarding publication channel

| Publication channel | # Occurrences | Distribution (%) |
|---|---|---|
| Conference article | 27 | 46 |
| Journal articles | 23 | 39 |
| Technical reports | 8 | 13 |
| Book chapter | 1 | 2 |

### 4.2  RQ-2. What Are the Publication Channels that Are Carried Out Most Frequently in Relation to the Topic Raised?

In Table 4, there is a list with the number of studies grouped by the Publication channel. The highest number of occurrences is the articles from conferences and journals with 85% of the total approximately. This high number is caused by the execution of search string in recognized digital databases due to these data sources collect academic researches, which are published mainly in specialized conferences or journals.

Additionally, 46% of total studies were published in journals, where 34% correspond to journals from a Medical context, while 43% were obtained from journals related to Software Engineering context. In Table 4, we could see the percentage of studies regarding research source.

Also, the distribution of articles found on journals correspond Medical, Software Engineering and other contexts such as Sociology and Education. It is important to mention that there are multiple studies found in Medical sources due to the SLR is a research methodology widely used in that context.

### 4.3  RQ-3. What Are the Difficulties Found in the Development of an SLR?

In the Table 5, we display the difficulties identified in the elaboration of a SLR. Take into account that we obtained the list from the extraction of different contexts. The

**Table 5.** Difficulties identified on the elaboration of SLR

| Difficulties | Articles | # Occurrences |
|---|---|---|
| Definition of research question | D2, D10, D13, D16, D18, D20, D23, D25, D30, D31, D46, D46, D50, D59 | 14 |
| Background review | D7, D9, D11, D21, D39 | 5 |
| Build search string | D4, D5, D6, D15, D16, D28, D31, D33, D35, D38, D40, D46, D49, D50, D54, D56 | 16 |
| Selection of primary studies | D2, D16, D20, D25, D29, D32, D40, D46, D47, D49, D55, D57 | 12 |
| Accessibility to data sources | D1, D2, D28, D38 | 4 |
| Validation with experts | D2, D16, D20, D21, D54 | 5 |

difficulties found on working on SLR are focused on the first activities of SLR. For instance, "Definition of research question" and "Build search string" activities. Those activities require multiple iterations and, in several cases, they require a considerable amount of time that affects dramatically in the execution of an efficient SLR. Also, we identified that "Selection of primary studies" process is described as a though task that also has negative impact on the results obtained. In the studies [D9], [D18], the wrong selection of studies impact on the quality of the SLR regardless if the methodology was implemented correctly.

### 4.4    RQ-4. What Are the Methodologies, Protocols or Similar Found for Conducting an SLR?

In the Table 6, we display the proposals found for working on SLR. We divided the proposals in five types: Methodology, Guide, Framework, Protocol and Model in order to classify and group the proposals. There are different proposals for working on SLR.

**Table 6.** Methodologies, protocols and guides found for conducting SLR

| Proposal | Type | ID |
|---|---|---|
| PRISMA | Methodology | D2 |
| Borrego, Foster, and Froyd Guide for SLR | Guide | D3 |
| Three states framework (Input/Processing/Output) | Framework | D9 |
| M Arksey H, O'Malley L Methodology | Methodology | D10 |
| EBSE lightweight model based on Kitchenham and Dybå | Guide | D15 |
| Template Sections Model | Model | D16 |
| Protocol of the "Center for Reviews and Dissemination (CRD)" | Protocol | D18 |
| SLR Delaware Guide | Guide | D23 |
| Enfoque iSR | Model | D31 |
| Methodi Ordinatio | Methodology | D55 |
| ProKnow-C | Methodology | D55 |
| Selection Methodology | Methodology | D57 |

From the results, we found that "Methodology" is the most used category among the proposals. On the contrary, protocol and framework are the least used.

### 4.5   RQ-5. What Are the Techniques and Tools Used in the Elaboration of a SLR?

In this question, we have considered a deep approach on SLR to get techniques and tools. In the Table 7, we present the tools or techniques for each proposal. There are multiple proposals on the primary studies identified. In order to get additional information of the distinct proposals, we also included the "Purpose" of the technique or tool to identify the main SLR activity that has the attention of researchers. The result obtained in Table 7 could be explained due to Selection activity. The Selection activity involves manual job that requires a high amount of time. As a result, the academy is working on getting tools and techniques for reducing it without affecting the quality of the expected results.

**Table 7.** Techniques and tools found on SLR preparation

| Name | Type | Purpose | ID |
|------|------|---------|-----|
| VisNave (Visualization and Navigation) | Tool | Extract | D19 |
| CiteSeerX | Tool | Selection | D19 |
| GoPubMed | Tool | Selection | D19 |
| CircleView | Tool | Selection | D19 |
| EEECA (Examine, Evaluate, Establish, Compare and Argue) | Technique | Extract | D23 |
| SQ3R (Survey, Question, Read, Write and Review) | Technique | Extract | D23 |
| Visual Text Mining | Technique | Selection | D32 |
| Approach "Skoglund and Runeson" | Technique | Search | D38 |
| Approach "Greenhalgh and Peacock" | Technique | Search | D38 |
| Citation Matrix | Technique | Selection | D38 |
| Score Citation Automatic Selection | Technique | Selection | D49 |
| StArt (State of the Art through Systematic Review) | Tool | Selection | D49 |
| Revis (Syst. Literature Review based on Visual Analytics) | Tool | Selection | D49 |
| Projection Explorer (PEx) | Tool | Results | D51 |

### 4.6   RQ-6. What Are the Techniques or Tools Used in the Search Strings Build?

In the Table 8, we show proposals categorized as Techniques and Tools used in the build of search strings.

There is an interest on defining multiple alternatives to build search strings. In fact, some of the techniques proposed were used on SLR in order to demonstrate the accuracy of the generated search strings. We found also that the proposed search strings require the evaluation of researchers for getting the final version.

**Table 8.** Tools and techniques used on building search strings

| Name | Type | ID |
|------|------|-----|
| De Montfort's Keywords | Technique | D23 |
| MICROARRAS | Tool | D11 |
| CANSEARCH | Tool | D11 |
| RUBRIC | Tool | D11 |
| PLEXUS | Tool | D11 |
| OCLC | Tool | D11 |
| Backward/Forward Search | Technique | D9 |
| "Start Set" definition | Technique | D38 |
| Backward/Forward Snowballing | Technique | D38 |
| Indeterminacy of language (Alternative Terminology) | Technique | D47 |
| Ovid Technologies searching system | Tool | D48 |
| PICO/PICOS/SPIDER | Tool | D58 |

In addition, we found that building the search strings requires more alternatives on techniques to define more efficient tools due to the automation of this process has a high impact on the execution time of SLRs. Finally, we have identified that there are research methods that present tools to complement the build and use of search strings.

### 4.7   RQ-7. What Are the Techniques and Tools Used as Search Strategies in a SLR?

On this question, the objective is to find recommendations for getting the ideal primary studies for the review. These recommendations were grouped as search strategies, it involves techniques and tools to identify articles and reduce research scope. In Table 9, we display the techniques and tools found.

There are more proposed techniques on search strategies. In fact, several of the techniques are related to the process of Selection. Due to this is an important activity on the SLR elaboration, the academy is involving on getting more efficient ways of identifying studies that will help the elaboration of SLR.

### 4.8   RQ-8. What Are the Techniques and Tools Used to Evaluate the Quality of Studies in a SLR?

Due to the SLR is a process that involves a well-defined protocol, it requires the review of the final studies found for guarantee a valid knowledge in the research. As a result, there are multiple techniques and tools implemented on the academic community that help researches on doing the quality assessment. In Table 10, we display the techniques and tools found.

**Table 9.** Techniques and tools used as search strategies in a SLR

| Name | Type | ID |
|---|---|---|
| Scoping Review | Technique | D2 |
| Snowball | Technique | D2 |
| Contacting Experts | Technique | D2 |
| Hand Searching | Technique | D2 |
| "29 techniques for information search"- Search Formulation Tactics | Technique | D7 |
| TREAD (Search strategies to find, evaluate, synthesize) | Tool | D14 |
| Keyword relationship | Technique | D19 |
| Visualization of Interrelations | Technique | D19 |
| ID3 Algorithm | Tool | D24 |
| Precision-based strategies | Technique | D26 |
| Quasi-Gold Standard | Technique | D33 |
| Snowball based on references (Skoglund and Runeson) | Technique | D34 |
| Search strategies for articles | Technique | D41 |

**Table 10.** Techniques and tools used in quality assessment of SLR articles

| Name | Type | ID |
|---|---|---|
| DESMET | Tool | D5 |
| Peer-review Process | Technique | D9 |
| CASP (Critical Appraisal Skills Programmed) | Tool | D14 |
| AGREE | Technique | D14 |
| GRADE | Technique | D29/D50 |
| DARE | Technique | D29 |
| MOOSE | Technique | D57 |
| Threshold approach | Technique | D57 |

# 5   Threats, Conclusion and Future Work

During the SLM, there were validations in the planning and the methodology used. Other members of the project performed these validations. However, despite peer review and assurance of the methodological framework, we have considered situations and activities that can influence on the results and conclusions obtained.

The main threat that was identified is the selection bias, because the primary study selection has a direct consequence on the research results.

On the other hand, we could conclude that there are plenty interest of the academy for getting better practices, techniques, methodologies, even tools for performing SLR on Software Engineering contexts. In fact, we found that there are multiple approaches from different contexts that we could adapt to Software Engineering environment to perform more efficient and accurate SLRs.

From the Medicine, Education and Sociology contexts, we could extract interesting guides and techniques for doing both "Studies Selection" and "Research Question" elaboration. These approaches could complement the current guidelines we found on SLR for Software Engineering. In addition, we could review the difficulties on SLR elaboration in order to visualize how these items could impact in the researches.

Finally, the methodologies, techniques, tools and guides found will help us to improve the current guidelines applied on SLR elaboration in software contexts. As research scope, we have considered "Search String", "Search Strategies", "Quality Assessment" and "Protocols and Methodologies in SLR elaboration". We consider we could expand the research on getting also "Threats to Validity" and "Recommendations on SLR elaborations".

**Acknowledgements.** This work is framed within the "Soporte a RSL" project funded by "III Fondo Concursable para la Innovación en la Docencia Universitaria" by Dirección Académica del Profesorado and partially supported by the Grupo de Investigación y Desarrollo de Ingeniería de Software (GIDIS) from the Pontificia Universidad Católica del Perú.

## Appendix A. Studies Included in the Review

[D1]  Cobo, M., López-Herrera, A., Herrera-Viedma, E., & Herrera, F. Science Mapping Software Tools: Review, Analysis, and Cooperative Study Among Tools. *Journal of the Association for Information Science and Technology, 67*(7), 1382–1402. (2011).

[D2]  Borrego, M., Foster, M., & Froyd, J. Systematic Literature Reviews in Engineering Education and other Developing Interdisciplinary Fields. *Journal of Engineering Education, 103*(1), 45–76. (2014).

[D3]  Godwin, A. Visualizing Systematic Literature Reviews to Identify New Areas of Research. *Frontiers in Education Conference (FIE)* (págs. 1–8). IEEE. (2016).

[D4]  Marshall, C., & Brereton, P. Tools to Support Systematic Literature Reviews in Software Engineering: A Mapping Study. *Empirical Software Engineering and Measurement* (págs. 296–299). IEEE. (2013).

[D5]  Marshall, C., Brereton, P., & Kitchenham, B. Tools to Support Systematic Reviews in Software Engineering: A Feature Analysis. *Proceedings of the 18th International Conference on Evaluation and Assessment in Software Engineering* (pág. 13). ACM. (2014).

[D6]  Bailey, J., Zhang, C., Budgen, D., Turner, M., & Charters, S. Search Engine Overlaps: Do they agree or disagree? *Realising Evidence-Based Software Engineering* (pág. 2). IEEE. (2007).

[D7]  Sherman, S., Bohler, J., & Shehane, R. Improving Student Literature Research Skills. *Global Education Journal, 2016*(1), 18–31. (2016).

[D8]  Adams, J., Hillier-Brown, F., Moore, H., Lake, A., Araujo-Soares, V., White, M., & Summerbell, C. Searching and Synthesising 'grey literature'and 'grey information' in Public Health: Critical Reflections on Three Case Studies. *Systematic Reviews, 5*(1), 164. (2016).

[D9]    Levy, Y., & Timothy, E. A Systems Approach to Conduct an Effective Literature Review in Support of Information Systems Research. *Informing Science Journal, 9*, 171–181. (2006).

[D10]   Kastner, M., Tricco, A., Soobiah, C., Lillie, E., Perrier, L., Horsley, T., & Straus, S. What is the Most Appropriate Knowledge Synthesis method to Conduct a Review? Protocol for a Scoping Review. *BMC Medical Research Methodology, 12*(1), 114. (2012).

[D11]   Gauch, S., & Smith, J. Query Reformulation Strategies for an Intelligent Search Intermediary. *AI Systems in Government Conference* (págs. 65–71). IEEE. (1989).

[D12]   Torraco, R. Writing Integrative Literature Reviews: Guidelines and Examples. *Human resource development review, 4*(3), 353–367. (2005).

[D13]   Kitchenham, B., Brereton, P., Budgen, D., Turner, M., Bailey, J., & Linkman, S. Systematic Literature Reviews in Software Engineering – A Systematic Literature Review. *Information and Software Technology*, 7–15. (2009).

[D14]   Krainovich-Miller, B., Haber, J., Yost, J., & Kaplan, S. Evidence-based Practice Challenge: Teaching Critical Appraisal of Systematic Reviews and Clinical Practice Guidelines to Graduate Students. *Journal of Nursing Education, 48*(4), 186–195. (2009).

[D15]   Rainer, A., & Beecham, S. *Supplementary Guidelines, Assessment Scheme and evidence-based evaluations of the use of Evidence Based Software Engineering.* University of Hertfordshire. (2008).

[D16]   Biolchini, J., Mian, P., Natali, A., & Travassos, G. *Systematic Review in Software Engineering.* System Engineering and Computer Science Department COPPE/UFRJ. (2005).

[D17]   Carliner, S. Workshop in Conducting Integrative Literature Reviews. *Professional Communication Conference (IPCC)* (págs. 1–3). IEEE. (2011).

[D18]   Boaz, A., Ashby, D., & Young, K. *Systematic Rreviews: What have they got to Offer Evidence Based Policy and Practice?* London: ESRC UK Centre for Evidence Based Policy and Practice. (2002).

[D19]   Saad, F., Mathiak, B., & Mutschke, P. Supporting Literature Review by Searching, Visualizing and Navigating Related Papers. *Cloud and Green Computing (CGC)* (págs. 363–368). IEEE. (2013).

[D20]   Kitchenham, B. *Procedures for Performing Systematic Reviews.* Keele University, UK. (2004).

[D21]   Reed, L. Performing a Literature Review. *Frontiers in Education Conference (FIE)* (págs. 380–383). IEEE. (1998).

[D22]   Coppola, N., & Carliner, S. Is our peer-reviewed literature sustainable? *Professional Communication Conference (IPCC)* (págs. 1–7). IEEE. (2011).

[D23]   Okoli, C., & Schabram, K. *A Guide to Conducting a Systematic Literature Review of Information Systems Research.* (2010).

[D24]  Rúbio, T., & Gulo, C. Enhancing Academic Literature Review through Relevance Recommendation. *11th Iberian Conference on Information Systems and Technologies*, (págs. 15–18). Gran Canaria, Spain. (2016).

[D25]  Hart, C. *Doing a Literature Review: Releasing the Social Science Research Imagination.* London: SAGE. (1998).

[D26]  Dieste, O., & Padua, A. Developing Search Strategies for Detecting Relevant Experiments for Systematic Reviews. *Empirical Software Engineering and Measurement* (págs. 215–224). IEEE. (2007).

[D27]  Laghrabli, S., Benabbou, L., & Berrado, A. A New Methodology for Literature Review Analysis Using Association Rules Mining. *Intelligent Systems: Theories and Applications (SITA)* (págs. 1–6). IEEE. (2015).

[D28]  Singh, P., & Singh, K. Exploring Automatic Search in Digital Libraries–A Caution Guide for Systematic Reviewers. *Proceedings of the 21st International Conference on Evaluation and Assessment in Software Engineering* (págs. 236–241). ACM. (2017).

[D29]  Zhang, H., & Babar, M. Systematic Reviews in Software Engineering: An Empirical Investigation. *Information and Software Technology, 55*(7), 1341–1354. (2013).

[D30]  Kitchenham, B., Brereton, P., Li, Z., Budgen, D., & Burn, A. Repeatability of Systematic Literature Reviews. In Evaluation & Assessment in Software Engineering. *Evaluation & Assessment in Software Engineering (EASE)*, (págs. 46–55). (2011).

[D31]  Lavallee, M., Robillard, P., & Mirsalari, R. Performing Systematic Literature Reviews with Novices: An Iterative Approach. *Transactions on Education, 57*(3), 175–181. (2014).

[D32]  Felizardo, K., Salleh, N., Martins, R., Mendes, E., MacDonell, S., & Maldonado, J. Using Visual Text Mining to Support the Study Selection Activity in Systematic Literature Reviews. *Empirical Software Engineering and Measurement (ESEM)*, (págs. 77–86). (2011).

[D33]  Zhou, X., Jin, Y., Zhang, H., Li, S., & Huang, X. A Map of Threats to Validity of Systematic Literature Reviews in Software Engineering. *Software Engineering Conference (APSEC)*, (págs. 153–160). (2016).

[D34]  Jalali, S., & Wohlin, C. Systematic Literature Studies: Database Searches vs. Backward Snowballing. *Proceedings of the ACM-IEEE international symposium on Empirical software engineering and measurement* (págs. 29–38). ACM. (2012).

[D35]  Carver, J., Hassler, E., Hernandes, E., & Kraft, N. Identifying Barriers to the Systematic Literature Review Process. *Empirical Software Engineering and Measurement* (págs. 203–212). IEEE. (2013).

[D36]  Aveyard, H. *Doing a Literature Review in Health and Social Care: A Practical Guide.* McGraw-Hill Education (UK). (2014).

[D37]  Reutzel, T. Outpatient Drug Insurance: a Framework to Guide Literature Review, Research, and Teaching. *American Journal of Pharmaceutical Education, 62*(1). (1998).

[D38] Wohlin, C. Guidelines for Snowballing in Systematic Literature Studies and a Replication in Software Engineering. *Proceedings of the 18th international conference on evaluation and assessment in software engineering* (pág. 38). ACM. (2014).

[D39] Wang, Y., Liu, D., Qu, H., Luo, Q., & Ma, X. A Guided Tour of Literature Review: Facilitating Academic Paper Reading with Narrative Visualization. *Proceedings of the 9th International Symposium on Visual Information Communication and Interaction*, (págs. 17–24). (2016).

[D40] Singh, P., & Singh, K. Exploring Automatic Search in Digital Libraries–A Caution Guide for Systematic Reviewers. *Proceedings of the 21st International Conference on Evaluation and Assessment in Software Engineering* (págs. 236–241). ACM. (2017).

[D41] Conn, V., Isaramalai, S., Rath, S., Jantarakupt, P., Wadhawan, R., & Dash, Y. Beyond MEDLINE for literature searches. *Journal of Nursing Scholarship, 35*(2), 117–182. (2003).

[D42] Yoshii, A., Plaut, D., McGraw, K., Anderson, M., & Wellik, K. Analysis of the Reporting of Search Strategies in Cochrane Systematic Reviews. *Journal of the Medical Library Association: JMLA, 97*(1), 21. (2009).

[D43] Im, E., & Chang, S. A Systematic Integrated Literature Review of Systematic Integrated Literature Reviews in Nursing. *Journal of Nursing Education, 51*(11), 632–636. (2012).

[D44] Wong, S., Wilczynski, N., & Haynes, B. Comparison of top-performing Search Strategies for Detecting Clinically Sound Treatment Studies and Systematic Reviews in MEDLINE and EMBASE. *Journal of the Medical Library Association, 94*(4), 451. (2006).

[D45] Sampson, M., McGowan, J., Cogo, E., & Horsley, T. Managing database overlap in systematic reviews using Batch Citation Matcher: case studies using Scopus. *Journal of the Medical Library Association, 94*(4), 461. (2006).

[D46] Riaz, M., Sulayman, M., Salleh, N., & Mendes, E. Experiences Conducting Systematic Reviews from Novices' Perspective. *Proceedings of the 14th international conference on Evaluation and Assessment in Software Engineering (EASE)*, (págs. 44–53). (2010).

[D47] Boell, S., & Cecez-Kecmanovic, D. On being 'Systematic' in Literature Reviews in IS. *Journal of Information Technology, 30*(2), 161–173. (2015).

[D48] Montori, V., Wilczynski, N., Morgan, D., & Haynes, B. *Optimal Search Strategies for Retrieving Systematic Reviews from Medline: Analytical Survey.* BMJ. (2005).

[D49] Octaviano, F., Felizardo, K., Maldonado, J., & Fabbri, S. Semi-Automatic Selection of Primary Studies in Systematic Literature Reviews: is it Reasonable? *Empirical Software Engineering and Measurement (ESEM), 20*(6), 1898–1917. (2015).

[D50] Kitchenham, B., Brereton, P., Turner, M., Niazi, M., Linkman, S., Pretorius, R., & Budgen, D. Refining the Systematic Literature Review Process —two participant-Observer Case Studies. *Empirical Software Engineering, 15*(6), 618–653. (2010).

[D51] Felizardo, K., Riaz, M., Sulayman, M., Mendes, E., MacDonell, S., & Maldonado, J. Analysing the Use of Graphs to Represent the Results of Systematic Reviews in Software Engineering. *Software Engineering (SBES)* (págs. 174–183). IEEE. (2011).

[D52] Santos, R., & Da Silva, F. Motivation to Perform Systematic Reviews and their Impact on Software Engineering Practice. *Empirical Software Engineering and Measurement* (págs. 292–295). IEEE. (2013).

[D53] Dieste, O., Anna, G., Juristo, N., & Saxena, H. Quantitative Determination of the Relationship Between Internal Validity and Bias in Software Engineering Experiments: Consequences for Systematic Literature Reviews. *Empirical Software Engineering and Measurement (ESEM)* (págs. 285–294). IEEE. (2011).

[D54] Babar, M., & Zhang, H. Systematic Literature Reviews in Software Engineering: Preliminary Results from Interviews With Researchers. *Empirical Software Engineering and Measurement* (págs. 346–355). IEEE. (2009).

[D55] Pagani, R., Kovaleski, J., & Resende, L. Methodi Ordinatio: A Proposed Methodology to Select and Rank Relevant Scientific Papers Encompassing the Impact Factor, Number of Citation, and Year of Publication. *Scientometrics, 105*(3), 2109–2135. (2015).

[D56] Ghafari, M., Saleh, M., & Ebrahimi, T. A Federated Search Approach to Facilitate Systematic Literature Review in Software Engineering. *International Journal of Software Engineering & Applications, 3*(2), 13. (2012).

[D57] Meline, T. *Selecting Studies for Systematic Review: Inclusion and Exclusion Criteria.* (2006).

[D58] Methley, A., Campbell, S., Chew-Graham, C., McNally, R., & Cheraghi-Sohi, S. PICO, PICOS and SPIDER: a comparison study of specificity and sensitivity in three search tools for qualitative systematic reviews. *BMC health services research, 14*(1), 579. (2014).

[D59] Oates, B., & Capper, G. Using Systematic Reviews and Evidence-Based Software Engineering with Masters Students. *Evaluation & Assessment in Software Engineering (EASE)*, (págs. 20–21). (2009).

[D60] Miranda, J., Muñoz, M., Uribe, E., Márquez, J., Uribe, G., & Valtierra, C. Systematic Review Tool to Support the Establishment of a Literature Review. *New Perspectives in Information Systems and Technologies*, (págs. 171–181). (2014).

# References

1. Kitchenham, B., Brereton, P., Turner, M., Niazi, M., Linkman, S., Pretorius, R., Budgen, D.: Refining the systematic literature review process—two participant-observer case studies. Empir. Softw. Eng. **15**(6), 618–653 (2010)

2. Reed, L.: performing a literature review. In: Frontiers in Education Conference (FIE), pp. 380–383. IEEE (1998)

3. Brereton, P., Kitchenham, B., Budgen, D., Khalil, M., Turner, M.: Lessons from applying the systematic literature review process within the software engineering domain. J. Syst. Softw. **80**(4), 571–583 (2007)
4. Lavallee, M., Robillard, P., Mirsalari, R.: Performing systematic literature reviews with novices: an iterative approach. Trans. Educ. **57**(3), 175–181 (2014)
5. Levy, Y., Timothy, E.: A systems approach to conduct an effective literature review in support of information systems research. Inf. Sci. J. **9**, 171–181 (2006)
6. Wohlin, C.: Guidelines for snowballing in systematic literature studies and a replication in software engineering. In: Proceedings of the 18th International Conference on Evaluation and Assessment in Software Engineering, p. 38. ACM (2014)
7. Marshall, C., Brereton, P.: Tools to support systematic literature reviews in software engineering: a mapping study. In: Empirical Software Engineering and Measurement, pp. 296–299. IEEE (2013)
8. Kitchenham, B., Brereton, P., Budgen, D., Turner, M., Bailey, J., Linkman, S.: Systematic literature reviews in software engineering – a systematic literature review. Inf. Softw. Technol. **51**(1), 7–15 (2009)
9. Kitchenham, B., Brereton, P., Li, Z., Budgen, D., Burn, A.: Repeatability of systematic literature reviews. In: Evaluation & Assessment in Software Engineering (EASE), pp. 46–55 (2011)
10. Borrego, M., Foster, M., Froyd, J.: Systematic literature reviews in engineering education and other developing interdisciplinary fields. J. Eng. Educ. **103**(1), 45–76 (2014)
11. Kitchenham, B., Charters, S.: Guidelines for Performing Systematic Literature Reviews in Software Engineering. Elsevier, Staffordshire (2007)
12. Ghafari, M., Saleh, M., Ebrahimi, T.: A federated search approach to facilitate systematic literature review in software engineering. Int. J. Softw. Eng. & Appl. **3**(2), 13 (2012)
13. Riaz, M., Sulayman, M., Salleh, N., Mendes, E.: Experiences conducting systematic reviews from novices' perspective. In: Proceedings of the 14th International Conference on Evaluation and Assessment in Software Engineering (EASE), pp. 44–53 (2010)
14. Carver, J., Hassler, E., Hernandes, E., Kraft, N.: Identifying barriers to the systematic literature review process. In: Empirical Software Engineering and Measurement, pp. 203–212. IEEE (2013)
15. Zhang, H., Babar, M.: Systematic reviews in software engineering: an empirical investigation. Inf. Softw. Technol. **55**(7), 1341–1354 (2013)
16. Petersen, K., Mujtaba, S., Feldt, R., Mattsson, M.: Systematic mapping studies in software engineering. 12th International Conference on Evaluation and Assessment in Software Engineering, pp. 1–10 (2008)
17. Barbara, K., Budgen, D., Brereton, P.: Using mapping studies as the basis for further research – a participant–observer case study. Inf. Softw. Technol. **53**(6), 638–651 (2011)
18. Dybå, T., Dingsøyr, T.: Empirical studies of agile software development: a systematic review. Inf. Softw. Technol. **50**(9–10), 833–859 (2008)
19. Shea, B., Grimshaw, J., Wells, G., Boers, M., Andersson, N., Hamel, C., Bouther, L.: Development of AMSTAR: a measurement tool to assess the methodological. BMC Med. Res. Methodol. (2007)
20. Popay, J., Arai, L., Rodgers, M., Birtten, N.: Guidance on the conduct of narrative synthesis in systematic reviews: a product from the ESRC methods programme (2006)
21. Jalali, S., Wohlin, C.: Systematic literature studies: database searches vs. backward snowballing. In: Proceedings of the ACM-IEEE International Symposium on Empirical Software Engineering and Measurement, pp. 29–38. ACM (2012)

# From Craftsmen into Engineers During Undergraduate Education

Eduardo Juárez[1]([⊠]), Edgar Fernández[1], José Velázquez[2],
Rocío Aldeco-Pérez[1], Lilia Rodríguez[2], Antonio Del Rio[3],
and Claud Robinson[4]

[1] School of Engineering and Science, Tecnologico de Monterrey,
Monterrey, Mexico
{edjuarezp, edfernand, raldeco}@itesm.mx
[2] School of Humanities and Education,
Epigmenio González 500, Querétaro, Mexico
{jmvelazq, lcrodrig}@itesm.mx
[3] Callpicker, Tecnológico Norte 801 Piso 4, Querétaro, Mexico
antonio@callpicker.com
[4] West Corporation, 11808 Miracle Hills Dr. Omaha, Omaha, NE, USA
claudr@gmail.com

**Abstract.** The software development industry has an urge for trained software engineers capable to deliver successful software projects. To be successful, software projects require engineers with discipline on the application of sound engineering practices. Those practices are usually learnt after engineers have spent years working in the industry, resulting in a low percentage of successful projects. In this paper we propose a challenge-based learning experience that develops such sound engineering practices and discipline during undergraduate studies, therefore, guaranteeing that the rate of successful projects where the engineers will participate will be higher, as they have acquired the competencies to deliver such projects.

**Keywords:** Challenge-based · Competency-based · Education
Undergraduate · Training · Agile

## 1 Introduction

The software development industry has had a steady growth rate over the last years and it is predicted that in the coming years, that growth rate will rise [1–3]. This growth creates an urge for trained software engineers capable to develop successful software projects [4]. It is documented that a large percentage of software projects exceed the delivery time agreed or even are never delivered [4]. This lack of success is usually consequence of inexistent sound engineering practices applied with discipline and responsibility [4]. The earlier those practices are learnt, the higher the percentage of successful projects will be [4]. Hence, the aim of this paper is to propose an undergrad-level challenge-based learning experience that develops an understanding and conscious capability to apply such sound engineering practices. Those practices are

© Springer Nature Switzerland AG 2019
J. Mejia et al. (Eds.): CIMPS 2018, AISC 865, pp. 31–40, 2019.
https://doi.org/10.1007/978-3-030-01171-0_3

included in the definition of competencies that later can be evaluated to verify that when present the project success rate is higher.

The structure of this paper as follows. In Sect. 2, the problem of efficient software development is presented as well as solutions adopted by the software industry and the academy. In Sect. 3, the evolution of the proposed learning experience is explained including improvements made to guarantee that the defined competencies are developed, and projects are successful. In Sect. 4, associated results are presented and explained. Finally, Sect. 5 outlines future work and offers some concluding remarks.

## 2   The Problem of Efficient Software Development

As stated in [4], large software projects are almost always over budget, usually delivered late, and, when delivered, are filled with defects. As many as 35% of large applications will be cancelled and never delivered at all.

According to [4], the software industry has the highest failure rate of any engineering field, spending around 50% of costs on fixing software defects. In [4], is also argued that, five common problems observed on unsuccessful projects are: (1) Estimates prior to starting the project are inaccurate and excessively optimistic; (2) Quality control during the project is poor; (3) Change control during the project is inadequate; (4) Tracking of progress during development time is severely misleading; and (5) Problems are ignored or concealed rather than dealt with rapidly and effectively when first noted.

At the organizational level, in [5] is reported that nearly 50% of organizations do not have standard processes, process assets, and job aids; 41% admit not to properly develop team members' individual skills and capabilities for future challenges; 42% have no established standard-planning process; and 54% do not measure what matters.

On the contrary, when successful projects are examined after completion and delivery, the implementation of standard practices in planning and estimation, quality control, change management, tracking progress and conflict resolution are found [4]. Therefore, it is clear that successful software projects tend to follow sound engineering practices, when unsuccessful ones do not.

Even though the problem has been identified, the percentage of successful software projects is still low. Then, software development industry has proposed some solutions mainly based on in-house training. Another important actor is the academy that has proposed to train students before they join the work force. In the following sections, an overview of proposed solutions from both actors is presented.

### 2.1   Industry

The software development industry [4] has created several approaches which have proven success on large software projects. Among these are the Capability Maturity Model Integration (CMMI) [6], the Team Software Process (TSP) [7] and Personal Software Process (PSP) [8]. The Rational Unified Process (RUP) [9] has also shown success. For smaller applications, variants of Agile development such as SCRUM [10] have proven also to be effective.

Overall, hybrid methods, such as the process decision framework Disciplined Agile Delivery (DAD) [11], have been also successful. The reason is that each of the previous methods in "pure" form has a rather narrow band of project sizes and types for which they are most effective. Combinations and hybrid methods are more flexible and match the characteristics of any size and type of project. However, care and expertise are required in putting together those methods to be sure that the best combinations are chosen. DAD is the first framework to provide guidance about finding such a combination. The previous statements are consistent with the 1987 declaration "No Silver Bullet" [12].

To put up with the methods, software development organizations need disciplined and skilled people. As stated by [4], the best software employers typically train their software engineers with an intensive onboarding training for about 4 to 10 weeks, an annual in-house training from 5 to 10 days per year, and 1 to 3 external commercial seminars per year, monthly or quarterly webinars on newer topics, among others. But this is not the case for most organizations, which some have criticized for their adoption of certification programs suggesting that they have opted for certificates rather than competence [13]. Therefore, industry usually expects academia to solve the problem by demanding graduates with true software engineering capabilities.

### 2.2 Academy

An important strength of undergraduate education is that what gets taught tends to be used throughout the rest of the professional lives of the students [4]. It takes an average of about three years of in-house training and on-the-job experience before a newly graduated software engineer could be entrusted with serious project responsibilities. This is about two years longer than the training period needed by electrical or mechanical engineers. As [4] concludes, software engineering curricula lagged traditional engineering curricula in teaching subjects of practical value. A quick review of several software engineering curricula found some serious gaps in academic training. Among the topics that seemingly went untaught were software cost estimating, design and code inspections, statistical quality control, maintenance of legacy applications, metrics and measurements, Six Sigma methods, risk and value analysis, function points, and joint applications design (JAD) for requirements analysis [4]. While basic technical topics are taught good at the university level, project management associated topics are far from state-of-the-art levels.

Currently the general focus in Academy is to include software engineering and project management courses in the curricula alongside with a project-oriented learning approach and sometimes including real clients. This is documented in [14–18], among others. However, in practice, software engineering and project management topics are usually isolated, or due to University and curricula restrictions fail to provide a true real experience with projects that require state-of-the-art engineering practices to be successful as stated in [4]. As observed in [16] and in the authors' experience, if students do not face real clients, they will be skeptical of software engineering practices because academy will not convey the experience and demands of production quality software development, and if academics act as uncertain customers, they will be rejected by the students as malevolent and implausible. Another problem is the focus of assessment, which it is typically focused on features delivered, rather than in responsibility for longer term technical debt.

As stated in [18], if software engineering is the main competence of a major or minor program, software engineering topics (e.g., process, requirements, design, development, testing, project management, quality assurance, ethics and soft skills) should be emphasized in the curriculum rather than programming.

In conclusion, graduated software engineers have an integral problem of discipline, including tools, processes and skills. At Tecnologico de Monterrey Campus Querétaro, we have faced this problem on the design of a software engineering experience offered to Computer Systems Engineering degree called "Consolidating an Information Technology Department".

## 3    Iterative Design of the Experience

Around 1998, at Tecnologico de Monterrey Campus Querétaro a learning experience that develops software engineering skills in order to meet the industry demands was designed. The learning experience has evolved since then up to these days with three major stages. The first stage called *The Software Engineering Block "El Bloque"*, followed by the *IT Department* and finally evolved as the *Semester i: Consolidating an IT Deparment*. On next subsections, each stage is described including its engineering and academic scenarios, the major pedagogical innovations and its evaluation system, and the produced results.

### 3.1    The Software Engineering Block *"El Bloque"*

In the first stage of the learning experience (from 1998 to 2016) a Learning–by–doing [19] approach was taken. This experience intended to develop software engineering skills to meet the industry demands and to graduate computer systems engineers with working experience. The learning experience simulates a software development team of 5 to 9 members in size. It was commonly called just *"El Bloque"* due to the integration of three courses into one in terms of grades and work to be done. The teams in *El Bloque* performed a software project with a real-world client and the project was the capstone project of the student's major.

This environment was appropriate for Competency Based Learning [20], however *El Bloque* was not yet formally declared as such, then competencies were not declared neither evaluated. The evaluation was focused on delivered products.

In the first years of *El Bloque*, the engineering guidance was sparse. Students tried to be successful by any means they could figure. Software Quality was viewed as a stage of testing before the end of the project. However, the Learning–by–doing experience proved to be so valuable which led to the creation of an organizational structure inside the Tecnologico de Monterrey called Development Support Center (CAD after its acronym in Spanish) which had the responsibility to provide to each major program a Learning–by–doing experience with real world clients.

Around 2007 the PMBOK [21] and CMMI [6] models were incorporated to the learning experience, moving the evaluation to not just products but to include process definition, execution and improvements, and Project Control. Years later, an iterative development approach was taken with the RUP one. The engineering teams designed their processes by adapting RUP to their own needs. The lecturers performed not only as teachers, but also as consultants of the students. The experience up to this point is presented in [22] along with the institutional challenges to execute and replicate the learning experience [23].

Between 2010 and 2014, 13 of 14 projects were accepted by the stakeholders on schedule and into scope. However, most of them were not within the cost that was measured using student time effort, which was estimated in 24 h per student per week (the academic units of *El Bloque*). Most of the projects had a CMMI level 2 with some practices of the CMMI level 3 such as verification and validation, process definition, process focus and risk management. When this CMMI results were shared to CMMI consultants from the local industry, they were surprised by such results, due to the 1-3 years period of a common enterprise to reach level 2. Even, one of the professors –on his first participation in the experience– did not believe it until the end of the semester when he validated the results.

On another side of the experience, the professors observed a need of support from experts in the development of communication, ethics, leadership, security and software architecture skills.

In 2015, teachers observed that RUP offers very little room to change or build upon. RUP, as defined by IBM, is a framework with a defined process, focused on deliverables. This means there are a set of steps and exits for each step to be followed. Therefore, professors decided to adopt DAD, which is a process decision framework. DAD allows teams to design, improve and own their software engineering process with a people first, learning oriented, agile, hybrid, IT solution focused, goal driven, delivery focused, enterprise aware, risk and value driven, and scalable approach.

## 3.2 IT Department

The second stage of El Bloque came into place in 2016 and 2017 after the educational model "*TEC21*"[1] of Tecnologico de Monterrey was announced. TEC21 incorporated the design of learning experiences based on Competency Based Learning and Challenge Based Learning [24], among others. One of this learning experiences is called "*Semester i*" consisting of a full-time immersion in a challenge, supported by a staff of professors who perform different roles, such as challenge designers, lecturers, tutors, mentors, evaluators and coordinators.

Before designing a complete immersion and to be prepared for it, the professors decided to incrementally improve the experience. The first change was to increase the real-world experience by not only simulating a software development team, but also an Information Technology Department composed by self-organized software development teams[2]. The challenge for the IT department was that all the stakeholders should be delighted by the end of their project, which should be finished and accepted in time,

---

[1] http://modelotec21.itesm.mx/.

[2] https://elbloque.org/.

cost and scope. Working as an IT department provided the students the flexibility, responsibility and autonomy to build and evolve a team capable of achieving this challenge. With these abilities, the IT department is able to select, from a redefined selection made by the professors, which projects will be built.

Based on the observation stated in Sect. 3.1, communication and ethics workshops provided by professors from the Humanities School were incorporated as a second change. Although, these workshops increased the defined skills, the levels reached did not comply with the ones defined on the graduate profile, therefore an improved strategy was required.

The third change was in the evaluation process. The professors adopted the SCAMPI [25] evaluation method to assess the maturity of the software engineering process.

After two terms of IT Departments, it was observed that working as an organization was highly challenging for the students. In the first term, the experience was successful including two accepted projects within time and scope, however, reaching agreements was difficult given the lack of a dedicated work room. In the second term, the lack of a dedicated space was still an issue, although, certain problems with ethical behavior were observed. At the end, one project failed, and one was accepted but challenged. Students, professors and academic leaders agreed in that a dedicated work room to practice software engineering was needed, which is consistent with the initial DAD suggestion for the strategy Organize Physical Environment in the Form Work Environment process goal. Another main concern was that although the design of the experience was appropriate for Competency Based Learning, the evaluation was not focused on the development of competencies.

### 3.3    Semester I: Consolidating an IT Department

The current stage of *El Bloque* started in the fall term of 2017. The 4 professors participating in the previous stage formed a team with 3 more professors and participated to design a *"Semester i"* experience. In this stage, Challenge Based Learning and Competency Based Learning are formalized, stated and evaluated.

The defined challenge for the experience is defined as *"To consolidate a highly qualified and internationally competitive Information Technology Department capable of managing and developing complex software development projects"*. To consider that an IT Department is consolidated, all the projects selected by the department should delight stakeholders and must be completed within time, scope and cost, with at least a process maturity level 2 of CMMI.

Based on the graduate profile of the corresponding students and the analysis presented on Sect. 2, a set of competencies to be developed in the challenge were defined. The competencies are the following: (1) To produce a software solution which has to satisfy a need of a client based on international standards such as the CMMI; (2) To create software engineering processes following frameworks such as Disciplined Agile and CMMI with at least a maturity level 2; (3) To be part of an Information Technology Department who shares a vision and a common way of work; (4) To control the execution of software engineering projects based on frameworks such as the PMBOK; (5) To elaborate software operation manuals to ensure that the final users can consume the solution; (6) To expose clearly and concisely her advances and achievements to the different stakeholders; (7) To resolve interest conflicts and ethical problems of her profession according to the ethics code of the ACM[3] and the code of ethics defined by the department; (8) To generate new ideas and to establish the necessary actions to implement the ideas evaluating its feasibility; and (9) To motivate her teammates by setting goals and taking action to allow the team to work effectively.

A classroom is designated to be the permanent department office allowing a full-time immersion to support the development of the defined competencies. Also new roles for the professors are set. The full roster of professor roles is: A coordinator to find and link development partners with the university, a coordinator of the learning experience, a tutor for the students, a project management lecturer and coach, a software engineering lecturer and coach, a software quality lecturer and coach, a software architecture development coach, a communication coach, an ethics coach, and an entrepreneurship and leadership lecturer and coach.

Professors can play one or more of the previous defined roles. To maintain background diversity, they have different affiliations including School of Engineering and Science, School of Humanities and Education, and School of Management and Entrepreneurship. Also, some of them are consultants and have been certified by Carnegie Mellon University and the Disciplined Agile Consortium. Some have also started their software companies and one is also an IT Director. All of them have at least a master's degree.

In the challenge, there is flexibility in what, when and how the learning happens. Traditional lectures evolved into learning modules which do not necessarily happen every week. Also, the modules duration and schedule are based on the syllabus topic and the challenge stage. Learning modules support the challenge and not the other way around, where usually a project is defined to achieve the objectives of a traditional class. Most of the learning modules occur in the first eight weeks of the experience, meanwhile in the other ten weeks the interactions between professors and students are more likely as consultancy and personal coaching. Moreover, as a part of the learning experience, each student is guided in a personal manner defining a personal challenge based upon their strengths, improvement areas and motivation.

## 4  Evaluation and Results

To evaluate the development of the aforementioned competencies, every student had at least two sessions with an evaluation committee formed by the most experienced professors in the corresponding area. The evaluation is as follows. (1) Before the evaluation session, the student reflects on her learning and achievements and gathers evidence of it. (2) The student presents the evidence to the evaluation committee. (3) The committee reviews the evidence. (4) The committee and the student had an interview about the learnings and achievements. (5) The committee agrees on the

---

[3] https://www.acm.org/about-acm/acm-code-of-ethics-and-professional-conduct.

development level of the competency, which can be: (a) There is no learning evidence; (b) There is evidence of learning but there is not achievement; (c) There is evidence of learning and achievement; (d) There is evidence of learning and achievement and also evidence of supporting team members to learn and/or achieve; and (e) There is evidence of learning, but it is used in an unethical way.

The evaluation does not have a fixed period, so students request it within the time frame of the learning experience. Also, students can request multiple evaluations if there is still time to develop the competencies. To accredit the full learning experience students should demonstrate at least level 2 of development in each competency. Levels 3 and 4 are considered very successful. Level 1 and 0 are considered failure as the expected level of mastery is not reached.

Until June of 2018, two executions of the stage 3 of the learning experience have taken place with a total of 39 students. In Table 1 it is summarized the number of students with their achieved development level. Levels 3 and 4 are presented together as they both are considered very successful and the student has demonstrated achievement. Level 0 is omitted because so far, no student has obtained that level.

**Table 1.** Development of competencies in 2 executions of the learning experience

| Competency | Level 1 | Level 2 | Levels 3 and 4 |
|---|---|---|---|
| To produce a software solution which has to satisfy a need of a client based on international standards such as the CMMI | 0 | 19 | 20 |
| To create software engineering processes following frameworks such as Disciplined Agile and CMMI with at least a maturity level 2 | 0 | 9 | 30 |
| To be part of an Information Technology Department who shares a vision and a common way of work | 1 | 5 | 33 |
| To control the execution of software engineering projects based on frameworks such as the PMBOK | 1 | 5 | 33 |
| To elaborate software operation manuals to ensure that the final users can consume the solution | 0 | 4 | 35 |
| To expose clearly and concisely her advances and achievements to the different stakeholders | 0 | 2 | 37 |
| To resolve interest conflicts and ethical problems of her profession according to the ethics code of the ACM and the code of ethics defined by the department | 0 | 6 | 33 |
| To generate new ideas and to establish the necessary actions to implement the ideas evaluating its feasibility | 0 | 1 | 38 |
| To motivate her teammates by setting goals and taking action to allow the team to work effectively | 0 | 0 | 39 |

During these executions, 5 projects were successfully developed and according to the stablished requirements defined by the sound practices. Therefore, we have shown that by implementing our proposed learning experience the defined competencies are achieved and the project success rate reached 100%.

## 5 Conclusions and Future Work

In this paper, we present an undergrad-level challenge-based learning experience called "Semester i: Consolidating an IT Department". This experience is focused on developing industry sound practices in the training of software engineers. In addition of developing hard skills, the experience also develops soft skills facilitating to undergraduates the acquisition of the defined competencies. Those competencies are later evaluated showing that, when they are present, the project success rate is higher. Therefore, we can conclude that this learning experience is a useful tool to train software engineers that can contribute in the success of software development industry.

As future work, to show the effectiveness of the evaluation process, a deeper analysis on the used evaluation tools is necessary. Moreover, to include data of former students' success on industry will guarantee that learnt competencies are still present. Also, more iterations of this experience are necessary to show that the success rate maintains or decreases when competencies are not developed.

**Acknowledgments.** The authors would like to thank the team involved in the designing, implementation and evolution of this learning experience including but not limited to students, professors, development partners, directors, colleagues and staff.

## References

1. Bridgwater, A.: The future for software in 2018. Forbes (2017). https://www.forbes.com/sites/adrianbridgwater/2017/12/18/the-future-for-software-in-2018/
2. Kutcher, E.: The reality of growth in the software industry. McKinsey & Company (2015). https://www.mckinsey.com/industries/high-tech/our-insights/the-reality-of-growth-in-the-software-industry
3. Software & Programming Industry Growth, CSIMarket (2018). https://csimarket.com/Industry/Industry_Growth.php?ind=1011
4. Jones, C.: Software Engineering Best Practices: Lessons from Successful Projects in the Top Companies, 1st edn. McGraw-Hill Inc., New York (2010)
5. CMMI Institute: Thriving in the age of disruption. CMMI Institute, Pittsburgh, PA, Technical report (2017). https://cmmiinstitute.com/resource-files/public/marketing/whitepapers/thriving-in-the-age-of-disruption
6. Chrissis, M.B., Konrad, M., Shrum, S.: CMMI for Development: Guidelines for Process Integration and Product Improvement, 3rd edn. Addison-Wesley Professional, Boston (2011)
7. Humphrey, W.S.: Introduction to the Team Software Process. Addison-Wesley Longman Ltd., Essex (2000)
8. Humphrey, W.: Psp(Sm): A Self-Improvement Process for Software Engineers, 1st edn. Addison-Wesley Professional, Boston (2005)

9. Kruchten, P.: The Rational Unified Process: An Introduction, 3rd edn. Addison-Wesley Longman Publishing Co., Inc., Boston (2003)
10. Schwaber, K., Beedle, M.: Agile Software Development with Scrum, 1st edn. Prentice Hall PTR, Upper Saddle River (2001)
11. Ambler, S.W., Lines, M.: Disciplined Agile Delivery: A Practitioner's Guide to Agile Software Delivery in the Enterprise, 1st edn. IBM Press, Indianapolis (2012)
12. Brooks, F.P.: No silver bullet essence and accidents of software engineering. Comput. **20**(4), 10–19 (1987). https://doi.org/10.1109/MC.1987.1663532
13. Fraser, S.D., Brooks, F.P., Fowler, M., Lopez, R., Namioka, A., Northrop, L., Parnas, D.L., Thomas, D.: ""No silver bullet" reloaded: retrospective on" essence and accidents of software engineering"." In: Companion to the 22nd ACM SIGPLAN Conference on Object-oriented Programming Systems and Applications Companion, ser. OOPSLA 2007, pp. 1026–1030. ACM, New York (2007). http://doi.acm.org/10.1145/1297846.1297973
14. Kuno, N., Nakajima, T.: Design and implementation of training course for software process improvement engineers. In: 2016 23rd Asia-Pacific Software Engineering Conference (APSEC), pp. 381–384 (2016)
15. Portela, C., Vasconcelos, A., Oliveira, S., Souza, M.: The use of industry training strategies in a software engineering course: an experience report. In: IEEE 30th Conference on Software Engineering Education and Training (CSEE&T), pp. 29–36 (2017)
16. Simpson R., Storer, T.: Experimenting with realism in software engineering team projects: an experience report. In: IEEE 30th Conference on Software Engineering Education and Training (CSEET), pp. 87–96 (2017)
17. Letouze, P., de Souza, J.I.M., Silva, V.M.D.: Generating software engineers by developing web systems: a project-based learning case study. In: IEEE 29th International Conference on Software Engineering Education and Training (CSEET), pp. 194–203 (2016)
18. Kulkarni, V., Scharff, C., Gotel, O.: From student to software engineer in the Indian it industry: a survey of training. In 23rd IEEE Conference on Software Engineering Education and Training, pp. 57–64 (2010)
19. Gibbs, G.: Learning by Doing: A Guide to Teaching and Learning Methods. FEU (1988)
20. Tobón, S., Prieto, J., Fraile, J.: Secuencias didácticas: aprendizaje y evaluación de competencias. Pearson, Mexico (2010)
21. Project Management Institute: A Guide to the Project Management Body of Knowledge: PMBOK Guide, 3a edn. PMI Global Standard, Evanston (2004)
22. Laborde, F., Juárez, E., Cortés, R.: Ambientes de aprendizaje basados en simulaciones controladas: El bloque de ingeniería de software. CIIE Revista del Congreso Internacional de Innovación Educativa **1**(1), 40–46 (2015)
23. Juárez, E., Cortés, R., Laborde, F.: Retos institucionales del modelo tec21 para garantizar el desarrollo de competencias de egreso. CIIE Revista del Congreso Internacional de Innovación Educativa **1**(1), 47–53 (2015)
24. Johnson, L.F., Smith, R.S., Smythe, J.T., Varon, R.K.: Challenge Based Learning: An Approach for Our Time. New Media Consortium, Austin (2009)
25. SCAMPI Upgrade Team: Appraisal requirements for cmmi version 1.3 (arc, v1.3). Software Engineering Institute, Carnegie Mellon University, Pittsburgh, PA, Tech. Rep. CMU/SEI-2011-TR-006 (2011). http://resources.sei.cmu.edu/library/assetview.cfm?AssetID=9959

# Gamification for Improving Software Project Management Processes: A Systematic Literature Review

Liliana Machuca-Villegas[1]($\boxtimes$) and Gloria Piedad Gasca-Hurtado[2]

[1] Universidad del Valle, Calle 13 # 100-00, 760032 Cali,
Valle del Cauca, Colombia
liliana.machuca@correounivalle.edu.co
[2] Universidad de Medellín, Carrera 87 No. 30-65, 50026 Medellín, Colombia
gpgasca@udem.edu.co

**Abstract.** This systematic literature review aims to (a) understand the current state of gamification as a strategy for improving processes associated with software project management and (b) define future gamification in the context of software project management. For these purposes, we identified and analyzed studies based on research interests, including software project management areas explored with gamification, gamification elements used in this context, research methods, and industry type. Our findings indicate a predominance of studies in project management areas related to integration, resources, and scoping. The most commonly reported research method is the solution proposal, and the most commonly used gamification element is the point system. Future research must focus on addressing unexplored project management areas, which can be intervened with gamification as an improvement strategy to facilitate the implementation of good practices that impact the success of software development projects.

**Keywords:** Gamification · Software project management
Systematic literature review · Software process improvement

## 1 Introduction

Recently, gamification has been construed as a potential strategy for improving the engagement and motivation of software engineers [1]. It has been studied as an alternative for establishing guides in collaborative work and for improving the interaction among professionals and the transfer of knowledge [A49]. In addition, gamification is being used as a strategy to develop leadership skills and effective communication among the members of the software development team [2].

The nature of software projects makes their management complex and demanding in comparison to the management of other types of non-software-related projects. Software are characterized as intangible products based on changing requirements and are created based on the cognitive processes of a development team. In Addition, the development process is characterized because each project is unique and faces certain management challenges, e.g., dependence on the communication and coordination

© Springer Nature Switzerland AG 2019
J. Mejia et al. (Eds.): CIMPS 2018, AISC 865, pp. 41–54, 2019.
https://doi.org/10.1007/978-3-030-01171-0_4

rules of the development team, planning and cost/time estimate accuracy, dependence on team members' skills, and generation of innovation and creativity to achieve the objectives of the project [3–5].

As a consequence, the challenges of managing software development projects are related to the factors that impact the quality and success of the projects. Some of these factors are categorized as technical and human factors as well as process, product, or technology factors [6].

In particular, the difficulties associated with the performance of software development team roles, the collaborative work, and the need to integrate and share knowledge and experiences, are some aspects influence the productivity and quality of work and thus the results and success of the project.

This article presents the results of a systematic literature review (SLR) whose objective is to review the state of the art in the application of gamification as an improvement strategy in software project management processes. The results are extended to the context of the gamification elements used in software project management areas and their impact on process improvement based on systematic mapping [7].

The rest of the article is structured as follows. Section 2 presents the supporting theoretical context for the review, Sect. 3 describes the methodology used for the SLR, Sect. 4 presents our results, Sect. 5 discusses these results, and Sect. 6 proposes conclusions and plans for future research.

## 2   Theory

### 2.1   Gamification

Gamification is a recent concept that can be defined as the use of game design elements in non-gaming contexts [8]. It seeks to incorporate gaming mechanics and elements in non-gaming environments to improve the engagement, motivation, and performance of users by making their tasks more attractive [1].

There are different contexts wherein gamification is being applied. Some of the most prominent ones are commerce, education, health, organizational systems, work, and innovation. In this way, gamification attempts to positively influence results at a psychological and behavioral user level [9].

Consequently, gamification is applied as a tool to encourage behavioral change and promote desired attitudes in many fields [10]. Furthermore, gamification seeks to use the selected gaming elements to influence changes in user behavior, such as increased engagement, enjoyment and motivation [11].

In software engineering, there are studies that evidence the application of gamification as a supporting alternative in the improvement of their processes, e.g., the proposal of a gamification framework for software process improvement initiatives [12], Gamiware, a gamification platform for the improvement of software processes [A9], and a framework for software engineering gamification [A6].

However, it should be noted that gamification is still a research area with preliminary advances, as reflected in current literature reviews. To date, the literature on gamification has focused mainly on the fields of education, crowdsourcing, and health

[9, 13]; most other fields have received limited attention. However, the interest in incorporating gamification into other domains continues to grow and more varied domains and perspectives are currently being researched [14, 15] [A18].

## 2.2  Software Project Management

As part of project management, processes focused on the planning, monitoring, and control of the project [16, 17] are commonly identified. At the Project Management Body of Knowledge (PMBOK) [18] level, 10 knowledge areas with associated project management processes are identified. In turn, these processes are framed in groups of initiation, planning, execution, monitoring, control, and closure processes. The project management areas are project integration management, project scope management, project schedule management, project cost management, project quality management, project resource management, project communications management, project risk management, project procurement management, and project stakeholder management.

Within the software engineering context, Sommerville [4] considers software project management as an area for proper management of institutional budget and time constraints.

From this same perspective, Pressman [16] views the management of a software development project as a set of management activities related to the planning, monitoring, and control of people as well as processes and events that occur as software evolve from a preliminary concept to a fully operational implementation.

In the same way, the Project Management Institute (PMI) extends the definition of software project management as "the application of knowledge, skills, tools, and techniques for project activities to meet the project requirements" [3, 18].

# 3  Materials and Methods

SLR, which is proposed by Kitchenham [19], is a methodology used in the development of this study. It is considered as a rigorous method for reviewing research results. Through SLR, we can identify, evaluate, and interpret all important research studies available for a particular research question, subject area, or phenomenon of interest [19]. For this particular SLR, we used a formal protocol described for software engineering [20].

For the execution of the systematic review, an observation window was defined covering the period between August 2017–December 2017 and February 2018–June 2018.

The objectives of this study are to (a) understand the current state of gamification as a strategy for improving processes associated with software project management and (b) provide suggestions for future research. More specifically, this article focuses on the following research questions:

RQ1:  Which software project management areas are currently being explored using gamification as a strategy for improvement?

RQ2:  Which gamification elements have been used in existing gamification works for software project management?

RQ3:     Which research methods are being applied in this context?
RQ4:     Which types of industries are using gamification in software project
         management?

The following search chains were used to perform the SLR, derived from the objective of the study, (gamification or gamifying) AND (software project or software project management OR project management OR agile project management).

For these purposes, we selected scientific databases related to computer science, e.g., ACM Digital Library, ScienceDirect (subject: computer science), IEEE Xplore, Scopus, Springer, and Engineering Village.

A total of 1930 studies were obtained, which were reviewed based on the inclusion and exclusion criteria, such as duplicate studies, studies that evidence the use of gamification in software project management, studies wherein the relationship between title, abstract, and keywords and search chain terms is identified, and studies in English, Spanish, or Portuguese.

A more detailed review was performed considering the relationship between title, abstract, and keywords and search chain terms as well as the availability of the information of the selected articles in alignment with the objectives of this SLR.

From the review described above, 49 base studies were extracted and used as the sample for this work. The data extracted from these studies and the results of the SLR are presented in the following section.

A schematic of the literature search and selection procedure is shown in Fig. 1.

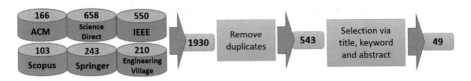

**Fig. 1.**  Study selection process.

Before identifying the relevant gamification aspects and determining the current status to define future lines of research, three questions are posed to assess the quality of these studies. The possible answers to these questions will respond whether the study meets, does not meet or partially meets the quality criteria associated with the questions. This assessment will provide information about the degree of gamification applied in software project management areas. The questions and their results are summarized in Table 1.

Finally, the extraction of information from each study was classified in a Microsoft Excel spreadsheet based on the following attributes: title, authors, country, continent, source, year, type of publication, keywords, abstract, research method used, management process, management area, contribution, gamification elements, gamification impact, and type of industry.

**Table 1.** Primary study assessment results.

| Nr. | Questions' assessment | Yes | No | Partially |
|-----|----------------------|-----|-----|-----------|
| QA1 | Does the study focus on the application of gamification in a software project management area? | 38 | 0 | 1 |
| QA2 | Does the study reflect the impact of gamification on aspects such as communication, engagement, collaboration, and motivation? | 19 | 16 | 4 |
| QA3 | Does the study reflect the impact of gamification on software development team's productivity? | 5 | 32 | 2 |

## 4 Results

With regard to the classification of primary studies according to their software project management areas, a comparative study of international reference models and standards focused on software project management was performed [21]. This study defines an equivalence between the different standards and models by considering the PMBOK model [18] as a reference for software [3]. The models compared against the reference model were CMMI-DEV [22] and SWEBOK [23]. Based on this comparison, we defined the basic management areas for the review. The distribution of the primary studies according to the established management areas is presented in Table 2.

**Table 2.** Distribution of primary studies according to project management areas.

| Management area | Quantity | Frequency (%) | Management area | Quantity | Frequency (%) |
|-----------------|----------|---------------|-----------------|----------|---------------|
| Integration | 16 | 32.7 | Resource | 19 | 38.8 |
| Scope | 9 | 18.4 | Communications | 0 | 0.0 |
| Schedule | 0 | 0.0 | Risk | 2 | 4.1 |
| Cost | 0 | 0.0 | Procurement | 0 | 0.0 |
| Quality | 3 | 6.1 | Stakeholder | 0 | 0.0 |

The studies were also classified according to the gamification elements used. Although there is not a commonly accepted taxonomy of gamification elements and mechanics [1], we used the elements identified by Pedreira et al. [1] and Muñoz et al. [A46]. Table 3 presents the classification of gamification elements according to the management areas. The table only includes elements whose frequency of use is greater than two.

To analyze how rigorous the research process was in each case, the primary studies have been classified based on their research method according to the proposal of Petersen et al. [24]: (i) validation research, (ii) evaluation research, (iii) solution proposal, (iv) philosophical papers, (v) opinion papers, and (vi) experience papers. Through this classification, we seek to identify the research approach used in the article. Table 4 shows the distribution of primary studies according to the type of research and management areas. Research method categories that did not include

**Table 3.** Distribution of primary studies according to project management areas and gamification elements.

| Gamification element | Management area | | | | | Total |
|---|---|---|---|---|---|---|
| | Integration | Resource | Scope | Quality | Risk | |
| Leaderboards | 3 | 3 | 4 | | | 10 |
| Point system/points | 4 | 3 | 7 | 1 | 2 | 17 |
| Badges | 4 | 2 | 4 | | | 10 |
| Levels | 4 | 2 | 2 | | | 8 |
| Progress bars | | | 3 | | 1 | 4 |
| Reward | 3 | 5 | 4 | | 1 | 13 |
| Scores | | | 2 | | 1 | 3 |
| Challenges | 1 | 1 | 2 | | 1 | 5 |
| Achievements | 2 | 2 | | | | 4 |
| Feedback | 2 | 1 | | | 1 | 4 |
| Dashboard | 3 | | | | | 3 |
| Ranking | 1 | 1 | 2 | | | 4 |
| Serious game | 3 | 5 | | 2 | | 10 |
| Quests | 1 | 2 | | | | 3 |
| Avatar | 1 | 1 | 1 | | | 3 |

**Table 4.** Distribution of primary studies according to the project management area and research method.

| Area | Validation | Solution Proposal | Philosophical | Experience | Total |
|---|---|---|---|---|---|
| Integration | [A6] | [A7–A11, A13–A17] | [A18–A21] | [A12] | 16 |
| Scope | [A23, A28, A30] | [A22, A24–A27, A29] | | | 9 |
| Quality | | [A1] | [A2, A3] | | 3 |
| Resource | [A34, A36, A37] | [A31–A33, A35, A41–A44, A47, A48] | [A38, A49, A45, A46] | [A39, A40] | 19 |
| Risk | | [A5, A4] | | | 2 |
| Total | 7 | 29 | 10 | 3 | 49 |

reference studies (evaluation research and opinion papers) and management areas wherein reference studies were not identified (cost, communications, schedule, procurement, and stakeholder) were eliminated.

The final study selection results were distributed in two types of environments according to the application of their contribution: (a) industry and (b) academics (Table 5).

**Table 5.** Distribution of selected primary studies by type of industry.

| Industry | Quantity | Frequency (%) | Industry | Quantity | Frequency (%) |
|---|---|---|---|---|---|
| Academics | 16 | 32.7 | Industry/academics | 6 | 12.2 |
| Industry | 13 | 26.5 | Unknown | 14 | 28.6 |

## 5  Discussion

The findings are expressed in response to the questions posed.

### 5.1  RQ1: Which Software Project Management Areas Are Currently Being Explored Using Gamification as a Strategy for Improvement?

According to the comparative study of models and international reference standards focused on software project management [21], half of the areas identified as project management areas evidence intervention with a gamification approach (Table 2).

The project resource management area stands out with 38.8%. The studies classified in this area were focused on promoting collaboration among team members, teamwork, leadership, and training.

In addition, the integration management area stands out with 32.7% and the project scope management area stands out with 18.4%. This process is characterized because it is a key process that requires active participation from stakeholders and influences the success of the project [18].

The project quality management area stands out with 6.1%, these studies emphasize on software testing and process improvement. In the case of the project risk management area, the percentage of studies classified was 4.1%, these studies emphasize on gamification strategies at an education or training level.

In conclusion, the software project management areas that explored gamification more as an improvement strategy mainly propose improvement strategies, e.g., (a) teamwork conditions, (b) interaction between stakeholders, and (c) participation of team members in a software development project.

Some of the studies identified are geared toward improvement in the requirement elicitation process. These studies promote the participation and engagement of stakeholders as well as improvements in the interaction and performance of team members.

However, there still remain unexplored management areas, including schedule management, cost management, communications management, procurement management, and stakeholder management.

Software project management is still an area of current research. Gamification as an improvement strategy can contribute to a project's success since it aims to promote collaboration and competitiveness among people, thereby improving their performance [A18].

### 5.2  RQ2: Which Gamification Elements Have Been Used in Existing Gamification Works for Software Project Management?

Table 3 lists 15 outstanding elements according to their frequency of use. The point system is the predominant element because it influences five management areas. The table of positions, badges, levels, rewards, and serious game also display predominance.

According to Hernández et al. [A49], the table of positions, badges, and point systems improve results and user participation and significantly contribute in the areas of scope management, resources, and integration. The quests and levels facilitate user motivation and orientation [A31]. On the contrary, the tables of positions facilitate feedback for the adoption of code navigation practices [A36].

With regard to the table of positions [A30], there is evidence of their contribution in meeting the requirements associated with status and power. In addition, challenges are suitable for people who demand curiosity and independence. In other studies, it was considered player types when assigning the gamification elements using Bartle's taxonomy [A5, A6, A9, A26]. However, the use of gamification elements at the business level still requires research and experimentation [A20].

As a result of the use of these gamification elements in software project management areas, a positive impact geared toward improving their processes is observed. In these studies, we observed an improvement in the participation of users in requirement elicitation as well as in collaboration, engagement, motivation, and communication between work team members. There is evidence of the use of gamification for team education or training, to improve productivity, contribute to the generation of a shared context and awareness of distributed teams, or stimulate and motivate the development team in the construction of knowledge.

### 5.3   RQ3: Which Research Methods Are Being Applied in this Context?

Table 4 presents a classification of the primary studies according to the research method used [24]. The solution proposal type occupies first class, with 59.2% of the primary studies presenting proposals for solving project management problems by implementing gamification strategies. These studies propose a specific tool, framework, method, or model for these purposes, with tools being the type of contribution that stands out the most. This category also includes games, platforms, software, or software prototypes.

The philosophical method (20.4%) involves group studies, e.g., (a) the state of the art of related topics, (b) exploratory studies, (c) literature reviews, or (d) systematic mappings.

The methods research validation (14.3%) and experience (6.1%) refer to other research studies. No related studies were found in the evaluation and opinion research method types.

Considering the result of this classification and assuming that the type of publication that predominates the most is conference, we can be concluded that most studies related to this topic are preliminary investigations that require new proposals for solving problems in this area of knowledge.

### 5.4   RQ4: Which Types of Industries Are Using Gamification in Software Project Management?

According to the type of contribution defined, in the academic environment, studies are classified at a higher percentage, i.e., 32.7%, compared to 26.5%, which corresponds to studies with contribution to the industry (Table 5). We were not able to classify 28.6% of the studies since the type of the industry in which they contribute is not explicit. On the contrary, six of the works were classified in both categories, industry and academics, due to their contribution.

Of the studies classified in the academic environment, the project resource management area occupies 37.5% of these studies, integration management occupies 43.75%, project risk management occupies 12.5%, and quality management occupies 6.25%.

With regard to the studies classified in the industry, the project management areas that stand out are integration management with 38.46%, scope management with 23.07%, and project resource management with 38.46%, with requirement elicitation being one of the most explored topics.

For the studies classified in the both the industry and academics contexts, 16.66% focus on integration management, 50% focus on scope management, and 33.33% focus on project resource management.

Based on these findings, we can conclude that in academics, the study of project resource management prevails, specifically with topics related to the work team. In addition, regarding the industry, the focus is mainly on project scope management, especially on requirement elicitation.

In summary, there is a growing interest in the implementation of gamification as a process improvement strategy in software project management. The increase in the number of primary studies determines that software project management is an area of research that is susceptible to the delimitation of future work lines.

Gamification has a positive impact on the development of skills, such as communication, participation, engagement, and motivation (QA2).

## 6  Conclusions and Future Work

In this article, an SLR was performed to characterize the state of the art regarding the application of gamification in software project management. Once the primary studies were selected, they were classified into different categories, which enabled their analysis and discussion based on the trend of the studies according to management areas, type of research, gamification elements, and type of industry.

The results of this SLR reflect the growing use of gamification and suggest it as a support alternative for the improvement of areas related to software project management. In addition, they favor the identification of future lines of work within this context. Some of these lines can be geared toward the following future works:

(a)  Analysis of the impact of gamification in software project management areas
(b)  Design of solution proposals wherein gamification can be systematically implemented in industrial environments
(c)  Design of gamification measurement models that demonstrate the efficiency in their implementation in software development environments
(d)  Identification of the relationship of the social and human factors with software development team's productivity.

# Appendix: List of All Studies Included Herein

A1.  Soska, A., Mottok, J., Wolff, C.: Pattern Oriented Card Game Development. 1170–1177 (2017).

A2.  Fraser, G.: Gamification of software testing. Proc. - 2017 IEEE/ACM 12th Int. Work. Autom. Softw. Testing, AST 2017. 2–7 (2017).

A3.  Jovanovic, M., Mesquida, A., Mas, A.: Process Improvement with Retrospective Gaming in Agile Software Development. 543, 287–294 (2015).

A4.  Uyaguari, F.U., Intriago, M., Jácome, E.S.: Gamification Proposal for a Software Engineering Risk Management Course. Adv. Intell. Syst. Comput. 353, III–IV (2015).

A5.  Moreta, L.L., Gamboa, A.C., Palacios, M.G.: Implementing a Gamified application for a Risk Management course. 2016 IEEE Ecuador Tech. Chapters Meet. 1–6 (2016).

A6.  García, F., Pedreira, O., Piattini, M., Cerdeira-Pena, A., Penabad, M.: A framework for gamification in software engineering. J. Syst. Softw. 132, 21–40 (2017).

A7.  Sharma, V.S., Kaulgud, V., Duraisamy, P.: A gamification approach for distributed agile delivery. Proc. 5th Int. Work. Games Softw. Eng. - GAS'16. 42–45 (2016).

A8.  Jurado, J.L., Fernandez, A., Collazos, C.A.: Applying gamification in the context of knowledge management. Proc. 15th Int. Conf. Knowl. Technol. Data-driven Bus. - i-KNOW'15. 1, 1–4 (2015).

A9.  Herranz, E., Colomo-Palacios, R., Seco, A. de A.: Gamiware: A Gamification Platform for Software Process Improvement. Commun. Comput. Inf. Sci. 425, 13–24 (2014).

A10. Ašeriškis, D., Damaševičius, R.: Gamification of a Project Management System. Seventh Int. Conf. Adv. Comput. Interact. Gamification. 200–207 (2014).

A11. Sharma, V.S., Kaulgud, V.: Agile workbench: Tying people, process, and tools in distributed agile delivery. Proc. - 11th IEEE Int. Conf. Glob. Softw. Eng. ICGSE 2016. 69–73 (2016).

A12. Sammut, R., Seychell, D., Attard, N.: Gamification of Project Management within a Corporate Environment. Proc. 15th Int. Acad. MindTrek Conf. Envisioning Futur. Media Environ. - MindTrek'11. 9 (2011).

A13. Parizi, R.M.: On the gamification of human-centric traceability tasks in software testing and coding. 2016 IEEE/ACIS 14th Int. Conf. Softw. Eng. Res. Manag. Appl. SERA 2016. 193–200 (2016).

A14. Passos, E.B., Medeiros, D.B., Neto, P.A.S., Clua, E.W.G.: Turning real-world software development into a game. Brazilian Symp. Games Digit. Entertain. SBGAMES. 260–269 (2011).

A15. Maxim, B.R., Kaur, R., Apzynski, C., Edwards, D., Evans, E.: An agile software engineering process improvement game. 2016 IEEE Front. Educ. Conf. 1–4 (2016).

A16. Gasca-hurtado, G.P., Gómez-Alvarez, M.C., Muñoz, M., Mejía, J.: Gamification Proposal for Defect Tracking in Software Development Process Gloria. In: Communications in Computer and Information Science. pp. 212–224 (2016).

A17. Mesquida, A., Jovanovic, M., Mas, A.: Process Improving by Playing: Implementing Best Practices through Business Games. 633, 225–233 (2016).

A18. Pedreira, O., García, F., Brisaboa, N., Piattini, M.: Gamification in software engineering - A systematic mapping. Inf. Softw. Technol. 57, 157–168 (2015).

A19. Souza, M.R.D.A., Veado, L.F., Moreira, R.T., Figueiredo, E.M.L., Costa, H. A.X.: Games for learning: Bridging game-related education methods to software engineering knowledge areas. Proc. - 2017 IEEE/ACM 39th Int. Conf. Softw. Eng. Softw. Eng. Educ. Track, ICSE-SEET 2017. 170–179 (2017).

A20. Olgun, S., Yilmaz, M., Clarke, P.M., O'Connor, R. V.: A Systematic Investigation into the Use of Game Elements in the Context of Software Business Landscapes: A Systematic Literature Review. In: International Conference on Software Process Improvement and Capability Determination. pp. 384–398 (2017).

A21. Kosa, M., Yilmaz, M.: Designing Games for Improving the Software. 1, 303–310 (2015).

A22. Kumar, B.S., Krishnamurthi, I.: Improving User Participation in Requirement Elicitation and Analysis by Applying Gamification Using Architect's Use Case Diagram. 49, 471–482 (2016).

A23. Busetta, P., Kifetew, F.M., Munante, D., Perini, A., Siena, A., Susi, A.: Tool-Supported Collaborative Requirements Prioritisation. Proc. - Int. Comput. Softw. Appl. Conf. 1, 180–189 (2017).

A24. Snijders, R., Dalpiaz, F.: Crowd-Centric Requirements Engineering. 0–1 (2014).

A25. Snijders, R., Dalpiaz, F., Brinkkemper, S., Hosseini, M., Ali, R., Özüm, A.: REfine: A gamified platform for participatory requirements engineering. 1st Int. Work. Crowd-Based Requir. Eng. CrowdRE 2015 - Proc. 1–6 (2015).

A26. Piras, L., Giorgini, P., Mylopoulos, J.: Acceptance Requirements and Their Gamification Solutions. Proc. - 2016 IEEE 24th Int. Requir. Eng. Conf. RE 2016. 365–370 (2016).

A27. Unkelos-Shpigel, N., Hadar, I.: Inviting everyone to play: Gamifying collaborative requirements engineering. 5th Int. Work. Empir. Requir. Eng. Emp. 2015 - Proc. 13–16 (2016).

A28. Ribeiro, C., Farinha, C., Pereira, J., Mira da Silva, M.: Gamifying requirement elicitation: Practical implications and outcomes in improving stakeholders collaboration. Entertain. Comput. 5, 335–345 (2014).

A29. Fernandes, J., Duarte, D., Ribeiro, C., Farinha, C., Pereira, J.M., Da Silva, M.M.: IThink : A game-based approach towards improving collaboration and participation in requirement elicitation. Procedia Comput. Sci. 15, 66–77 (2012).

A30.  Lombriser, P., Dalpiaz, F., Lucassen, G., Brinkkemper, S.: Gamified requirements engineering: Model and experimentation. Lect. Notes Comput. Sci. (including Subser. Lect. Notes Artif. Intell. Lect. Notes Bioinformatics). 9619, 171–187 (2016).

A31.  Diniz, G.C., Silva, M.A.G., Gerosa, M.A., Steinmacher, I.: Using gamification to orient and motivate students to contribute to oss projects. Proc. - 2017 IEEE/ACM 10th Int. Work. Coop. Hum. Asp. Softw. Eng. CHASE 2017. 36–42 (2017).

A32.  Steffens, F., Marczak, S., Filho, F.F., Treude, C., De Souza, C.R.B.: A preliminary evaluation of a gamification framework to jump start collaboration behavior change. Proc. - 2017 IEEE/ACM 10th Int. Work. Coop. Hum. Asp. Softw. Eng. CHASE 2017. 90–91 (2017).

A33.  Hof, S., Kropp, M., Landolt, M.: Use of Gamification to Teach Agile Values and Collaboration. Proc. 2017 ACM Conf. Innov. Technol. Comput. Sci. Educ. - ITiCSE'17. 323–328 (2017).

A34.  Prause, C.R., Jarke, M.: Gamification for Enforcing Coding Conventions. Proceeding 10th Jt. Meet. Eur. Softw. Eng. Conf. ACM SIGSOFT Symp. Found. Softw. Eng. (ESEC/FSE 2015). 649–660 (2015).

A35.  Sukale, R., Pfaff, M.S.: QuoDocs: Improving developer engagement in software documentation through gamification. Proc. Ext. Abstr. 32nd Annu. ACM Conf. Hum. factors Comput. Syst. - CHI EA'14. 1531–1536 (2014).

A36.  Snipes, W., Nair, A.R., Murphy-Hill, E.: Experiences gamifying developer adoption of practices and tools. Companion Proc. 36th Int. Conf. Softw. Eng. - ICSE Companion 2014. 105–114 (2014).

A37.  Yilmaz, M., Connor, R.V.O.: A Scrumban integrated gamification approach to guide software process improvement: a Turkish case study. Teh. Vjesn. - Tech. Gaz. 23, 237–245 (2016).

A38.  Hernandez, L., Munoz, M., Mejia, J., Pena, A., Rangel, N., Torres, C.: Application of gamification elements in software engineering teamwork [Aplicación de elementos de gamificación en equipos de trabajo en la ingeniería de software]. Iber. Conf. Inf. Syst. Technol. Cist. (2017).

A39.  Schafer, U.: Training scrum with gamification: Lessons learned after two teaching periods. IEEE Glob. Eng. Educ. Conf. EDUCON. 754–761 (2017).

A40.  Akpolat, B.S., Slany, W.: Enhancing Software Engineering Student Team Engagement in a High-Intensity Extreme Programming Course using Gamification. 149–153.

A41.  Butgereit, L.: Gamifying mobile micro-learning for continuing education in a corporate IT environment. 2016 IST-Africa Conf. IST-Africa 2016. 1–7 (2016).

A42.  Calderón, A., Ruiz, M., O'Connor, R. V: ProDecAdmin: A Game Scenario Design Tool for Software Project Management Training. In: European Conference on Software Process Improvement. pp. 241–248 (2017).

A43.  Calderón, A., Ruiz, M., O'Connor, R. V: Coverage of ISO/IEC 29110 Project Management Process of Basic Profile by a Serious Game. In: European Conference on Software Process Improvement. pp. 111–122 (2017).

A44. Muñoz, M., Hernández, L., Mejia, J., Peña, A., Rangel, N., Torres, C., Sauberer, G.: A Model to Integrate Highly Effective Teams for Software Development. In: European Conference on Software Process Improvement. pp. 613–626 (2017).

A45. Rangel, N., Torres, C., Peña, A., Muñoz, M., Mejia, J., Hernández, L.: Team Members' Interactive Styles Involved in the Software Development Process. In: European Conference on Software Process Improvement. pp. 675–685 (2017).

A46. Muñoz, M., Hernández, L., Mejia, J., Gasca-Hurtado, G.P., Gómez-Alvarez, M.C.: State of the Use of Gamification Elements in Software Development Teams. In: European Conference on Software Process Improvement. pp. 249–258 (2017).

A47. De Melo, A.A., Hinz, M., Scheibel, G., Diacui Medeiros Berkenbrock, C., Gasparini, I., Baldo, F.: Version control system gamification: A proposal to encourage the engagement of developers to collaborate in software projects. Lect. Notes Comput. Sci. (including Subser. Lect. Notes Artif. Intell. Lect. Notes Bioinformatics). 8531 LNCS, 550–558 (2014).

A48. Calderón, A., Ruiz, M.: Coverage of ISO/IEC 12207 Software Lifecycle Process by a Simulation-Based Serious Game. 155, 59–70 (2016).

A49. Hernández, L., Muñoz, M., Mejia, J., Peña, A.: Gamificación en equipos de trabajo en la ingeniería de software: Una revisión sistemática de la literatura Gamification in software engineering teamworks: A systematic literature review. Presented at the (2016).

# References

1. Pedreira, O., García, F., Brisaboa, N., Piattini, M.: Gamification in software engineering - a systematic mapping. Inf. Softw. Technol. **57**, 157–168 (2015). https://doi.org/10.1016/j.infsof.2014.08.007

2. Manrique-Losada, B., Gasca-Hurtado, G.P., Gomez Álvarez, M.C.: Assessment proposal of teaching and learning strategies in software process. Rev. Fac. Ing. 105–114 (2015). https://doi.org/10.17533/udea.redin.n77a13

3. Project Management Institute, IEEE Computer Society. Software extension to the PMBOK® Guide Fifth Edition. Project Management Institute, Inc. (2013)

4. Sommerville, I.: Software Engineering, 10th edn. (2016)

5. Ahmed, A.: Software project management (2012)

6. McConnell, S., Águila Cano, I.M., Bosch, A., et al.: Desarrollo y gestión de proyectos informáticos (1997)

7. Machuca-Villegas, L., Gasca-Hurtado, G.P.: Gamification for improving software project: systematic mapping in project management. In: 2018 13th Iberian Conference on Information Systems and Technologies (CISTI). IEEE (2018)

8. Deterding, S., Dixon, D., Khaled, R., Nacke, L.: From game design elements to gamefulness. Schr. zur soziotechnischen Integr. **3**(15), 2797 (2011). https://doi.org/10.1081/E-ELIS3-120043942

9. Hamari, J., Koivisto, J., Sarsa, H.: Does gamification work? - A literature review of empirical studies on gamification. Proc. Annu. Hawaii Int. Conf. Syst. Sci. 3025–3034 (2014). https://doi.org/10.1109/hicss.2014.377
10. Stieglitz, S., Lattemann, C., Robra-Bissantz, S., et al.: Gamification using game elements in serious contexts (2017)
11. Engedal, J.Ø.: Gamification - a study of motivational affordances. Dept. Comput. Sci. Media Technol. 1, 81 (2015)
12. Herranz, E., Colomo-palacios, R., Seco A de, A., Sánchez-Gordón, M.-L.: Towards a gamification framework for software process improvement initiatives: construction and validation. J. Univ. Comput. Sci. 22, 1509–1532 (2016)
13. Hamari, J.: Transforming homo economicus into homo ludens: a field experiment on gamification in a utilitarian peer-to-peer trading service. Electron. Commer. Res. Appl. 12, 236–245 (2013)
14. Heredia, A., Colombo-Palacios, R., Amescua-Seco, A.: A systematic mapping study on software process education, pp. 7–17 (2015)
15. Darejeh, A., Salim, S.S.: Gamification solutions to enhance software user engagement – a systematic review. Int. J. Hum. Comput. Interact. 7318(10447318), 1183330 (2016). https://doi.org/10.1080/10447318.2016.1183330
16. Pressman, R.S.: Software engineering a practitioner's approach 7th edn. Roger S. Pressman (2010)
17. Sommerville, I.: Ingenieria del Software - 7a edición. 687 (2005)
18. Project Management Institute: A Guide to the Project Management Body of Knowledge, 5th edn. Project Manager Instidute, Pennsylvania (2013)
19. Kitchenham, B., Charters, S.: Guidelines for performing systematic literature reviews in software engineering version 2.3. Eng. 45, 1051 (2007). https://doi.org/10.1145/1134285.1134500
20. Biolchini, J., Mian, P.G., Candida, A., Natali, C.: Systematic review in software engineering. Eng. 679, 1–31 (2005). https://doi.org/10.1007/978-3-540-70621-2
21. Clasificación de áreas de gestión de proyectos de acuerdo con los modelos de referencia PMBOK, CMMI-DEV y SWEBOK (2018)
22. CMMI Product Team.: Cmmi for development, version 1.3 (2010)
23. Bourque, P., Fairley, R.E. (Dick): Guide to the software engineering body of knowledge version 3.0 (2014)
24. Petersen, K., Feldt, R., Mujtaba, S., Mattsson, M.: Systematic mapping studies in software engineering. In: 12th International Conference on Evaluation and Assessment in Software Engineering (EASE 2008), pp. 1–10 (2008)

# Information Technology Service Management Processes for Very Small Organization: A Proposed Model

Freddy Aquino[1(✉)], Diego Pacheco[1], Paula Angeleri[2],
Rosanna Janampa[3], Karin Melendez[3], and Abraham Dávila[3]

[1] Facultad de Ciencias e Ingeniería, Pontificia Universidad Católica del Perú,
Lima, Peru
{freddy.aquino, dpachecov}@pucp.edu.pe
[2] Facultad de Ingeniería y Tecnología Informática, Universidad de Belgrano,
Buenos Aires, Argentina
paula.angeleri@comunidad.ub.edu.ar
[3] Departamento de Ingeniería, Pontificia Universidad Católica del Perú,
Lima, Peru
{janampa.rdp, kmelendez, abraham.davila}@pucp.edu.pe

**Abstract.** Nowadays Information Technology Services Management (ITSM) has become strongly needed for every kind of organizations providing IT services for customers or for themselves. However, existing models (as CMMI-SVC, ITIL or ISO/IEC 20000) are strongly difficult to implement on very small organizations. The aim of this article is to propose an ITSM model which can be applied in very small organization. Our methodology was: define an ITSM model considering needs and constraint in small organization, map elements to ISO/IEC 20000, and validate Model in small enterprises. The ITSM Model obtained was defined using relevance base practices from ISO/IEC 20000 and considering main characteristics of small organizations. The model proposed was validated in three small companies, with positive results.

**Keywords:** Information Technology Services Management (ITSM)
Small organization · ISO/IEC 20000 · ISO/IEC 29110

## 1 Introduction

Information Technology Services Management (ITSM) helps organizations to per-form operations and decision-making process [1]. Such organizations may be Information Technology (IT) services companies or IT departments that provide IT services for the entire organization, for the purpose of this article they will be considered IT service "entities". Every IT service entity, independently of its size (small, medium or large) must deliver its service in an ordered manner, in an improving environment culture in order to provide services with the required quality for satisfying its customers and accomplished business goals [2].

To help organizations to provide IT services with quality, some IT service models, framework or standards were developed, such as: ITIL [3], CMMI-SVC [4] o ISO/IEC

© Springer Nature Switzerland AG 2019
J. Mejia et al. (Eds.): CIMPS 2018, AISC 865, pp. 55–68, 2019.
https://doi.org/10.1007/978-3-030-01171-0_5

20000 [5]. Unfortunately, in many cases the adoption of these standards have been partial [6], for instance at [7] they focused mainly in incident management process implementation, at [8] incident management, problem management and service level agreement processes have been implemented, among others; in both cases [7, 8] the type of the implementation project was driven by the need of organizations to solve specific problems. According to [9] the adoption of ITSM models is difficult because of the complexity of the project and the lack of guidelines that support the proper introduction of these models in organizations. Particularly, this situation became critical among small organizations, for instance [9] stated that 52% of European Small Enterprises know these models, but only the 10% have implemented one or more of these standards.

The ProCal-ProSer project [10] has focused on Very Small Entities (VSEs). Its objectives were: to investigate the state of the art regarding the implementation of ITSM by the VSEs, develop an ITSM model for VSE that fits their needs, validate the model with implementation projects in VSEs and determine the key factors that influence the adoption of the proposed model. This project has adopted VSE definition from ISO/IEC 29110 [11] which states that a VSE is "an entity (enterprise, organization, department or project) having up to 25 people".

This article presents a proposed ITSM Model (under name PCPS4SVC) that can help VSEs to improve progressively processes and services. The article is organized in 6 sections: Sect. 2 presents ITSM models and selected works, Sect. 3 presents the protocol used for doing a proposed model, Sect. 4 describes the proposed ITSM model for VSE, Sect. 5 shows the implementation of the proposed ITSM model in a VSE, and Sect. 6 is a summary of the conclusions about the model and its implementation, and proposes future work.

## 2   Referential Framework

The referential framework for the proposed model is based on ITSM models (ITIL, CMMI-SVC) and standards (ISO/IEC 20000 series and ISO/IEC 15504-8), and some related articles. This section include: ISO/IEC models for ITSM, ITIL model, CMMI-SVC model, ISO/IEC 29110 software and systems engineering models for VSEs, and other authors selected works.

### 2.1   ISO/IEC Standards Referred to ITSM

The following ISO/IEC series of standards were considered for this project: ISO/IEC 20000 and ISO/IEC 15504:

- ISO/IEC 20000 is a series of standards, addressing Information Technology Service Management (ITSM) requirements and guidelines. Part 1 [5] established the requirements for an ITSM with the purpose of claiming conformance. It also includes definition of terms (vocabulary) that are used in other parts of this series of standards [5]. Part 4 defines a Process Reference Model (PRM) which establishes a set of processes for ITSM that is conformant with requirements specified in

ISO/IEC 20000-1 [12]. PRM for ITSM is composed by a set of 6 main processes: (i) general processes for ITSM System; (ii) processes for design and transition of new or changed services; (iii) processes for delivering services; (iv) processes for the resolution of incidents and problems; (v) processes for managing relationships with customers and suppliers; and, (vi) control processes that provide support to other processes. Part 4, established the elements needed for defining processes in conformance with PRM and with ISO/IEC 15504-2 [12]. There are other parts of this series of standards, but the proposed model presented in this article focused on parts 1 and 4, as they are the most relevant for establishing an ITSM model.

- ISO/IEC 15504 is a series of standards used for improving and assessment process capability and organizational maturity [13, 14]. This article presents only parts 2 and 8 because they are the most relevant standards for the proposed model. The ISO/IEC 15504-2 [14] establishes a set of requirements for models to be assessed, such as a ITSM Process Reference Model (PRM) ISO/IEC 20000-4 [12]. It also establishes evaluation requirements that composed the Process Assessment Model (PAM), described in ISO/IEC 15504-8 [14]. This standard establishes, for each process, purpose, outcomes, base practices, inputs and outputs [14]. ISO/IEC 15504-8 is a Process Assessment Model which presents process indicators that are helpful for service providers (when implementing an ITSM) and useful for services assessors when performing an assessment of an ITSM in conformance with ISO/IEC 20000-4 [12]. This standard presents two indicators: (i) process capability indicators such as generic practices, generic resources, generic inputs and outputs, these indicators allow to determine the degree of adoption of the attributes that define a level of process capability; (ii) process performance indicators such as basic practices, inputs and outputs of process that allow to give the degree of obtaining the results on the purpose of each process.

Nowadays, ISO/IEC 15504 series of standards are being replaced by ISO/IEC 330xx family of standards, in particular ISO/IEC 33020:2015 replaces ISO/IEC 15504-2 [15].

## 2.2 Information Technology Infrastructure Library - ITIL®

Information Technology Infrastructure Library (ITIL) was developed by the Central Computer and Telecommunications Agency at United Kingdom with the purpose of providing reliable and effective Information Technology (IT) services [16]. ITIL is a framework made to design, define, and maintain IT service management processes [17]. Many organizations have implemented ITIL, some have failed to implement it, others have succeeded but it has taken more time than they have planned; in summary ITIL implementation it is not an easy task [16].

ITIL best practices are described among 5 publications [17]: (i) ITIL Service Strategy describes how to define the most adequate strategy for managing IT services, with the purpose of achieving better effectiveness and profit; (ii) ITIL Service Design, based on strategies Service Design is essential for satisfying business need; (iii) ITIL-Service Transition, presents the mechanisms for the deployment and release of new services, or release of services that had major changes; (iv) ITIL-Service Operation, IT services operation must be managed in order to be conformant with IT Service Level

Agreements (SLAs), avoiding services interruptions, managing service incidents and solving problems when needed; (v) ITIL-Service continuous improvement, allows finding opportunities for making improvements on service management life cycle, throughout services measurement and evaluation.

## 2.3   CMMI® for Services

CMMI® for Services [4]: (i) is a model that provides guidelines for establishing and operating services, and providing services support; (ii) is a part of CMMI constellation shared process areas with CMMI for development and CMMI for acquisition (iii) is composed by 24 process areas (set of recommendations for improving performance in the area) classified in four categories: Project and work management, Establishing and providing services, Process management, and Support; (iv) has two representations: (a) continuous or by capacity levels, consisting of 4 levels (0–3) that apply to the process improvement achievements of an organization in a given process area; and (b) staged or maturity levels: it has 5 levels (1–5) and applies to the process improvement achievements of an organization with respect to a group of process areas; (v) where improvement achievements refer to generic goals (general purposes for all process areas) and specific goals (particular purposes for each process area); for this purpose, the model presents generic and specific practices that are recommendations whose application allows reaching the respective goals.

## 2.4   ISO/IEC 29110

According to ISO [18], micro, small and medium enterprises are present in most sectors and should benefit from the technical knowledge of the standards. There are many small and medium organizations in the world, but most of the ISO/IEC standards do not cover their needs [19]. In this contexts, in 2011 ISO published the first part of ISO/IEC 29110 series of standards, with the purpose of helping small organizations to improve the quality of their products and the performance of their processes [19, 20]. At that time, ISO/IEC 29110 presented four process profiles for VSEs, in Software and Systems domains [20] (i) Entry Profile: the organization, a start-up with less than three years of operations, or/and an organization that develops projects up to 6 months/person; (ii) Basic Profile: the organization develops only one project at a time by a single work team; (iii) Intermediate Profile: the organization develops multiple projects with more than one team; and (iv) Advanced Profile: The main idea is that the organization wants to grow and sustain itself as an independent competitive software and/or system development business. Also, in Part 3 [21], it is established that an evaluation scheme uses ISO/IEC 15504-2. In 2014 ISO members approved the study of a new standard addressing VSEs need of implementing an ITSM model. The new standard to be developed was assigned to working group WG24 of ISO/IEC Joint technical Committee Information Technology, subcommittee Software and Systems Engineering. A team was established, including two authors of this article, and an earlier version of this Proposed ITSM model was presented to the rest of team members. Recently, in January 2018, this new standard presenting an ITSM model for VSEs was published as ISO/IEC 29110-5-3 [22]. The proposed ITSM model of this article is aligned to it from its conception.

## 2.5    Related Works

The following are some cases of IT services process implementations in Very Small Entities (VSEs):

- Implementation of the management of the service catalogue [23]: the proposal was carried out in a private company, in which the services for the services catalogue were identified. The Company is focused on IT areas that provide services to other areas of IT. The authors saw the need to analyze IT services, perform evaluations of the costs of services, improve the definition of the offer, improve communication with customers and finally improve the planning and development of new or modified services. In order to respond to the need, the authors [23] developed a service management proposal, in which they described actions such as: (i) definition of the service; (ii) definition of service components; (iii) roles involved in service management; (iv) processes of identifying services according to business needs; and (v) service life cycle process.
- Construction of a catalogue of IT services in a small company as the main input of IT financial management [1]: according to the Proposal, standard categories of IT services are defined, and an IT service definition standard is defined for each category. The authors define the IT services catalogue model and then its evaluation model, considering the needs of the small business. The verification of the model was done using a case study in a small company with 18 employees and 5 areas [1]. Finally, the model represented a flexible proposal of standard categories and services, which can be used by other companies and also with the possibility of adding new categories.
- IT Service Management Maturity Model (MM-GSTI) applied to small and medium organizations [24]: this proposal, that seeks to meet the needs of GSTI, was created to be applicable in small and medium organizations that provide IT services. The proposal [24] can be used as a basis for generating other models, which is why it was used by the Brazilian Software Process Improvement Program (MPS.BR) for the creation of a service-oriented maturity model. The Proposal is compatible with ISO/IEC 20000, CMMI-SVC, MPS.BR and ITIL [24]. Its structure is based on MPS.BR and uses process capability concepts as defined in ISO/IEC 15504, establishing process attributes and results for each level of maturity. In the definition of processes, the model considered: processes and results compatible with ISO/IEC 20000 and CMMI-SVC v1.3. The main difference of the MM-GSTI model with others is that it presents 7 levels of maturity, which may be more suitable for adoption in small and medium enterprises [24]. The model was evaluated through expert surveys, obtaining its approval both in the definition of the processes and in the sequence of their adoption.

# 3   Protocol of the Work Done

The definition of the proposed Model was based on four major activities: (i) understand the small organizations that offer services, their needs, restrictions, current practices, etc.; (ii) adapt practices by selecting them from a base model; (iii) verify in a conceptual manner the alignment to another internationally accepted model; and, (iv) validation of the model in small organizations. The context and activities are presented below.

- As a part of the Competisoft Project in Peru [25–29], a software process improvement was performed, during six months, in twenty small software development companies (with less than 12 developers). The SPI was performed using MoProSoft as process model and ISO/IEC 15504-2 for process assessment [30]. In this context, it was observed that fourteen VSEs, offered attention to users to help solve the situations that arise in the products implemented by them. The developers themselves offered customer support in a disorderly manner, generating situations such as: duplication of work (two analysts solving the same problem without knowing that it was being resolved), and delay in development activities caused by their efforts to address the claims of customers, among others. These situations and tasks, in the literature, are considered as incidents and problems. During the Competisoft Project [30] it was established that adopting standards in small organizations is difficult, as indicated in the ISO guide [18], and that it is very difficult to adopt ITSM models in all kind of organizations in general [31], and in small organizations in particular [6]. The first step when developing the proposed ITSM model was the identification of common activities among ITSM standards and models, such as: ITIL 2011 [17], ISO/IEC 20000-1 [5], ISO/IEC 20000-4 [12], ISO/IEC 15504-8 [13], and CMMI-SVC 1.3 [4]. These common activities were compared and a common sequence of activities for ITSM models was established, in almost all cases. In addition, some complementary activities were also identified and taken into account for the development of the proposed model. Another step consisted in adopting the process pattern used in ISO/IEC 29110-5-1-2. Finally, the activities included in the proposed Model were determined by selecting base practices from ISO/IEC 15504-8 and CMMI-SVC, and reviewing them by looking for tailoring and consistency
- After the proposed Model was defined, a verification was performed using correspondence analysis and coverage analysis [32, 33]. In our case, a mapping of the processes elements of this proposed Model and its specifications was carried out, comparing them with the elements of the service management model described in ISO/IEC 15504-8 [13]. In our case, the main element considered was the activity of the analyst. The correspondence analysis [33] allows to determine the relationship between two elements; and the coverage analysis [33] allows to determine the degree to which an element of this proposed model covers a reference model (and vice versa). For this correspondence, a decomposition scheme was followed, in which one element (for example one activity) is decomposed into others of greater granularity in order to facilitate the identification of similar elements (or its decomposition) in the other model. For coverage, every element decomposed is

compared with the corresponded element (from the other model) and the degree of similarity is determined using rules and scale according to our references [33, 32]; the final coverage is determined by aggregation of partial coverages. In Sect. 5, presents PCPS4SVC model analysis.

- The proposed Model, in the first version, was used by three small organizations (enterprises) as a trial. The main activity of these enterprises is software development but they offer complementary IT services, such as providing support during their software implementation, deployment and maintenance.

## 4 A Proposed IT Service Management Model for VSEs

The PCPS4SVC model was developed to fulfill VSE need of adopting the existing ITSM practices. The proposal was based on the ISO/IEC 15504-8 (PAM of ISO/IEC 20000-4), ISO/IEC 29110 process pattern, experiences of authors, faculty and professionals regarding ITSM models. By the time PCPS4SVC model was published, ISO/IEC JTC1/SC7/WG24 was developing an ITSM model for VSEs and a first version of PCPS4SVC was presented to ISO/IEC consideration. The first version of PCPS4SVC model was implemented as a trial in three small enterprises and after that some adjusted was included in the Model proposed in this article.

Our Model includes two processes: Basic Service Management (BSM) and Service Operation Management (SOM). The BSM process was defined taking into account the ISO/IEC 15504 processes, and were identified as: CON.1 Change management, CON.2 Configuration management, REL.1 Business relationship management, SDE.6 Service level management, and SDE.7 Service reporting. Similarly, the SOM processes were defined as: CON.1 Change management, CON.2 Configuration management, RES.1 Incident management, RES.2 Service request management and RES.2 Problem management (see Fig. 1). The processes selected from ISO/IEC 15504 were based on two criteria: (i) processes that cover in a minimal way task related incidents and problems, called service operation; and, (ii) processes that articulate and stabilize the tasks of incidents and problems.

### 4.1 Basic Service Management (BSM) Process

- Purpose: To manage the customer's needs and expectations, making sure that the service satisfies the conditions agreed and signed between the IT service provider (VSE) and the customer.
- Objectives: O1 Establish communication between the service provider and the customer in order to identify its needs, expectations, claims, and satisfaction with the service provided; O2 Define the services to be provided in a service catalogue to inform the customer, the users and IT staff. This catalogue should always be available, accessible, and updated; O3 Define the service provided to a customer by means of a Service Level Agreement (SLA) documented and signed. The SLA defines the reach and the commitments agreed between the customer and the service provider with respect to the service to be provided; O4 Manage changes in the SLA

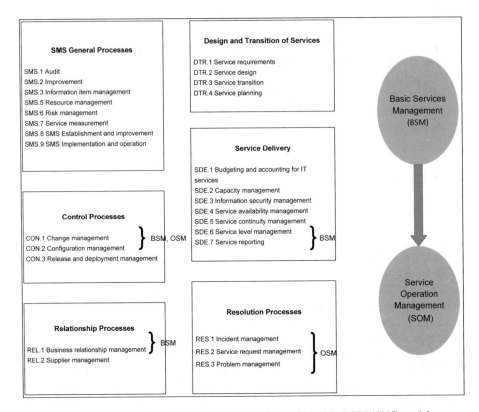

**Fig. 1.** Processes from ISO/IEC TS 15504-8 considered in PCPS4SVC model

originated by customer requests, quality improvements or temporary suspension of the service; O5 Inform all interested parties about the service performance and customer satisfaction.

- Activities Diagram: Fig. 2 shows the activities diagram and artifacts for the BSM process.

Appendix A shows a complete definition of Basic Service Management (BSM) Process.

## 4.2  Service Operation Management Process (SOM)

- Purpose: The purpose of the Service Operations Management (SOM) process is to ensure the operation of the service through the fulfilment of service requests, and resolution of incidents and problems that may affect the continuity of service operation.
- Objectives: O1 Meet the service request according to the established SLA, discriminating that request that is not within the scope of the service and/or that is not the competence of the service provider; O2 Restore the service according to the SLA in order to continue its operation; O3 Solve the problem from the investigation

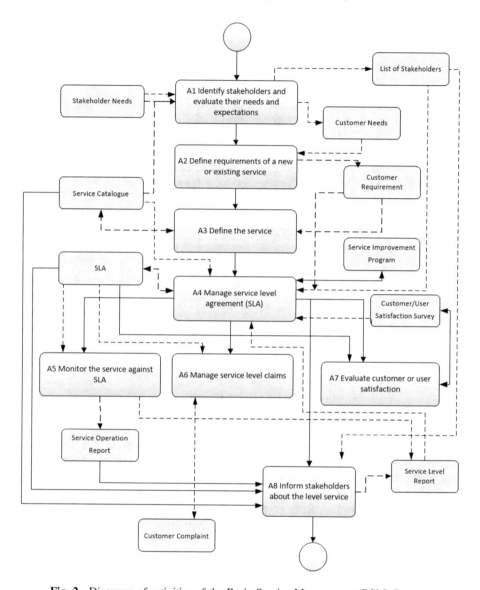

**Fig. 2.** Diagrams of activities of the Basic Service Management (BSM) Process

and identification of the root cause of the same, taking actions that allow to solve it and close it, avoiding that it is repeated again; O4 Consolidate the information of the operation of the service for an adequate decision making regarding the service.

• Activity diagram: Fig. 3 shows the activities and artifacts diagram for the Service Operation Management (SOM) process.

Appendix B shows a complete definition of Service Operations Management (SOM) process.

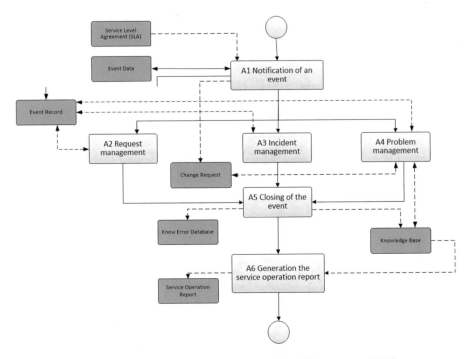

**Fig. 3.** Diagram of activities of the Service Operations Management (SOM) process

## 5   Verification and Trial of the PCPS4SVC Model

Section 3 describes the proposed Model verification using correspondence analysis and coverage analysis. Both techniques are explained and applied in other cases in [32, 33]. This verification was performed between PCPS4SVC model and ISO/IEC 15504-8.

Table 1 presents the result of the correspondence and coverage analysis of BSM process, which shows that 81% of the BSM process of the PCPS4SVC model is covered by the processes of ISO/IEC 15504-8. Similarly, the results for the SOM process are presented in Table 2. Both results showing that the proposed Model is aligned with ISO/IEC 20000 through ISO/IEC 15504-8.

During the Competisoft Project, where small software development companies participated, their processes for providing user support post software implementation were tracked and documented. From this information, process diagrams were constructed, selecting those that represent different implementations. In all cases, the diagrams obtained were mapped with respect to the PCPS4SVC model (version 1), and the proposed Model is conceived as an abstraction of them. This comparison confirmed that the proposed Model could be instantiated in different situations.

Finally, in the ProCal-ProSer Project context, some enterprises (three of them) accepted participate in a process improvement project using PCPS4SVC model. These enterprises have as a main activity software development however; they always offer the user support services under maintenance agreement on the software implemented.

**Table 1.** Coverage of the Basic Service Management (BSM) Process

| Elements | Elements of PCPS4SVC | Elements of ISO/IEC 15504 | Coverage (%) |
|---|---|---|---|
| Activity BSM.1 | 0.11 | 0.110 | 100 |
| Activity BSM.2 | 0.11 | 0.055 | 50 |
| Activity BSM.3 | 0.11 | 0.055 | 50 |
| Activity BSM.4 | 0.11 | 0.110 | 100 |
| Activity BSM.5 | 0.11 | 0.035 | 32 |
| Activity BSM.6 | 0.11 | 0.110 | 100 |
| Activity BSM.7 | 0.11 | 0.110 | 100 |
| Activity BSM.8 | 0.11 | 0.110 | 100 |
| Activity BSM.9 | 0.11 | 0.110 | 100 |
| Process coverage | 1.0 | 0.805 | 81 |

**Table 2.** Coverage of the Service Operations Management (SOM) process

| Elements | Elements of PCPS4SVC | Elements of ISO/IEC 15504 | Coverage (%) |
|---|---|---|---|
| Activity SOM.1 | 0.1 | 0.110 | 65 |
| Activity SOM.2 | 0.2 | 0.055 | 100 |
| Activity SOM.3 | 0.2 | 0.055 | 100 |
| Activity SOM.4 | 0.2 | 0.110 | 100 |
| Activity SOM.5 | 0.1 | 0.035 | 30 |
| Activity SOM.6 | 0.2 | 0.110 | 63 |
| Process coverage | 1.0 | 0.82 | 82 |

The adoption of the proposed Model was composed of the following macro-activities: (i) understand and document of the current situation identifying problems and business objectives; (ii) perform a diagnostic assessment regarding PCPS4SVC; (iii) make proposals to improve activities and deliverables based on the Model proposed; (iv) implement the improvement; and (v) perform a final assessment. The managers of the companies appreciated the model because they noticed how their user support service had improved. An article with more detail on the implementations made in VSEs has been written. This article will present the information in a comparative manner for a better analysis of the results.

## 6 Final Discussion and Future Work

The elaboration of the proposed ITSM model and its verification has allowed us to verify that the proposed activities and tasks are important for organizations of up to 25 people that provide IT services. The elaboration was based on existing models, professional experience and experiences from small organizations. The verification was carried out through a mapping of models that involved correspondence analysis and coverage analysis. The analysis was executed regarding the practices of ISO/IEC

20000-4 through the detailed elements of ISO/IEC 15504-8. The first version of the Model was introduced in three software development companies that offer IT support services to their customers. With their inputs, some adjustments were done on the Model presented in this article.

Although there are several model options for managing IT services, including some oriented towards small organizations, our proposal is also aligned to ISO/IEC 29110 series of standards. Our proposal was written as a guide of adoption for the basic profile and that consists of two processes covering mainly the operation processes. The coverage obtained showed that it is a lighter model that facilitates its adoption.

It is also planned to begin the elaboration of a process assessment model (PAM) based on ISO/IEC 15504-2 (or based on newest ISO/IEC 33000 standard series) for the proposed ITSM model.

**Acknowledgments.** This work has been carried out within the ProCal-ProSer project funded by Innóvate Perú under Contract 210-FINCYT-IA-2013 and partially supported by the Department of Engineering and the Research and Development Group of Software Engineering (GIDIS is the acronyms in Spanish) of Pontificia Universidad Católica del Perú.

## Appendixes

Appendix A and B are available in:
   https://drive.google.com/open?id=1aTxdjdV1Zy0YlPxt0BEXnxCxyRK3B9Jb.

## References

1. Arcilla, M., Calvo-Manzano, J.A., San Feliu, T.: Building an IT service catalog in a small company as the main input for the IT financial management. Comput. Stand. Interfaces **36** (1), 42–53 (2013)
2. Disterer, G.: Why firms seek ISO 20000 certification - a study of ISO 20000 adoption. Ecis, no. 2012, p. Paper 31 (2012)
3. ITIL: ITIL homepage (2012)
4. SEI: CMMI® for Services, Version 1.3 (2010)
5. ISO/IEC: ISO/IEC 20000-1 Information technology – Service managment – Part 1. Geneva (2011)
6. Melendez, K., Dávila, A., Pessôa, M.: Information technology service management models applied to medium and small organizations: a systematic literature review. Comput. Stand. Interfaces **47**, 120–127 (2016)
7. Marrone, M., Gacenga, F., Cater-Steel, A., Kolbe, L.: IT service management: a cross-national study of itil adoption. Commun. Assoc. Inf. Syst. **34**(49), 865–892 (2014)
8. Liu, M., Gao, Z., Luo, W., Wan, J. Case study on IT service management process. In: Evaluation Framework Based on ITIL, pp. 199–202. IEEE (2011)
9. Küller, P., Vogt, M., Hertweck, H.D., Grabowski, M.: A Domain specific IT service management approach for small & medium enterprises. In: 16th IBIMA conference on Innovation and Knowledge Management, Kuala Lumpur, vol. 2012, no. March 2016, pp. 1795–1807 (2011)

10. Dávila, A.: Proyecto [Pro]ductividad y [Cal]idad en [Pro]ductos software y [Ser]vicios software (ProCal-ProSer). GIDIS PUCP (2017). https://sites.google.com/a/pucp.pe/procal-proser/. Accessed 8 Aug 2017
11. ISO/IEC: ISO/IEC TR 29110-1:2011 Software engineering – Lifecycle profiles for Very Small Entities (VSEs) – Part 1: Overview. Geneva (2011)
12. ISO/IEC: ISO/IEC TR 20000-4 Information technology – Service managment – Part 4. Geneva (2010)
13. ISO/IEC: ISO/IEC TS 15504-8:2012 Information technology – Process assessment – Part 8: An exemplar process assessment model for IT service management. Geneva (2012)
14. ISO/IEC: ISO/IEC 15504-2:2003 Information technology – Process assessment – Part 2: Performing an assessment. Geneva (2003)
15. ISO: Migration ISO/IEC 15504 to ISO/IEC 33020:2015 (2018). https://www.iso.org/obp/ui/#iso:std:iso:9241:-11:ed-2:v1:en
16. Pereira, R., da Silva, M.M.: ITIL maturity model. In: 2010 5th Iberian Conference on Information Systems and Technologies (CISTI), pp. 1–6 (2010)
17. Axelos: axelos homepage (2001)
18. ISO: Guidance for writing standards taking into account micro, small and medium-sized enterprises' needs (2012)
19. ISO/IEC, "ISO/IEC TR 29110-5-1-2:2011 Software engineering – Lifecycle profiles for Very Small Entities (VSEs) – Part 5-1-2: Management and engineering guide: Generic profile group: Basic profile," Geneva (2011)
20. ISO/IEC: ISO/IEC 29110-2-1:2015 Software engineering – Lifecycle profiles for Very Small Entities (VSEs) – Part 2-1: Framework and taxonomy, Geneva (2015)
21. ISO/IEC: ISO/IEC TR 29110-3:2011 Software engineering – Lifecycle profiles for Very Small Entities (VSEs) – Part 3: Assessment guide, Geneva (2011)
22. ISO/IEC: ISO/IEC TR 29110-5-3:2018, "Systems and software engineering – Lifecycle profiles for Very Small Entities (VSEs) – Part 5-3: Service delivery guidelines", International Standardization Organization, Geneva, Switzerland (2018)
23. Mendes, C., Da Silva, M.M.: Implementing the service catalogue management. In: Proceedings of 7th International Conference on Quality of Information and Communications Technology, QUATIC 2010, pp. 159–164 (2010)
24. Machado, R.F., Reinehr, S., Malucelli, A.: Towards a Maturity Model for IT Service Management Applied to Small and Medium Enterprises, pp. 157–168 (2012)
25. Morillo, P., Vizcardo, M., Sanchez, V., Dávila, A.: Implementación y certificación de MoProSoft en una pequeña empresa desarrolladora de software: lecciones aprendidas de cuatro iteraciones de mejora. In: XI Simpósio Brasileiro de Qualidade de Software (SBQS), pp. 389–396 (2012)
26. Maidana, E., Vilchez, N., Vega, J., Dávila, A.: Identificación de problemas en proyectos de mejora de procesos : una experiencia en tres pequeñas empresas desarrolladoras de software en el Perú Resumen Introducción. In: VII Jornada Peruana de Computacion, pp. 120–129 (2008)
27. Mogrovejo, J., Dávila, A.: Una Experiencia de Implantación de COMPETISOFT en una Pequeña Empresa Desarrolladora de Software. In: VII Jornadas Iberoamericanas de Ingeniería de Software e Ingeniería del Conocimiento (JIISIC 2008), pp. 67–71 (2008)
28. Sánchez, G., Vergara, D., Dávila, A.: Experiencia de Implementación de Mejora de Procesos en dos PYME Desarrolladoras de Software, que poseen certificación ISO 9001 : 2000. In: VII Jornadas Iberoamericanas de Ingeniería de Software e Ingeniería del Conocimiento (JIISIC 2008), pp. 73–80 (2008)

29. Ñaupac, V., Arisaca, R., Dávila, A.: Software process improvement and certification of a small company using the NTP 291 100 (MoProSoft). In: Product-Focused Software Process Improvement. PROFES 2012. Lecture Notes in Computer Science, vol. 7343, pp. 32–43 (2012)
30. Dávila, A., et al.: The peruvian component of Competisoft project: lesson learned from academic perspective. In: 38th Latin America Conference on Informatics, CLEI 2012 (2012)
31. Melendez, K., Dávila, A.: Problemas en la adopción de modelos de gestión de servicios de tecnologías de información. Una revisión sistemática de la literatura. DYNA **85**(204), 215–222 (2018)
32. Alvarado, R., Delgado, L., Dávila. A.: Mapeo y evaluación de la cobertura de los procesos de MPS.Br a los procesos de la categoría de Operación de MoProSoft. In: XI Simpósio Brasileiro de Qualidade de Software (SBQS), pp. 158–172 (2012)
33. Canepa, K., Dávila, A.: Evaluación teórica de la capacidad de procesos de Rational Unified Process respecto del MoProSoft. Ind. Data **13**(2), 83–91 (2010)

# Exploring the Influence of Belbin's Roles in Software Measurement: A Controlled Experiment with Students

Raúl A. Aguilar[✉], Julio C. Díaz, Juan P. Ucán,
and Yasbedh O. Quiñones

Universidad Autónoma de Yucatán, Mérida 97000, Yucatán, Mexico
{avera, julio.diaz, juan.ucan}@correo.uady.mx,
sky_lab_beto@hotmail.com

**Abstract.** The article presents a controlled experiment in which the convenience of using the Belbin Role Theory for the integration of work teams with the task of measuring the software is explored. The study is developed in an academic environment with students of the Software Engineering degree and analyzes the differences between the metrics obtained with the Function Point Technique, by teams integrated with the Belbin Theory, and those obtained by teams integrated with students selected randomly. The results obtained provide evidence regarding the significant differences in terms of the time spent for the task; it was observed that teams integrated with the Belbin theory take more time. Regarding the five metrics obtained to measure the functionality of the software, differences were found only in the functionality linked to the external outputs.

**Keywords:** Software testing education
Computer-supported collaborative learning
Computer-supported cooperative work
Experimentation in software engineering

## 1 Introduction

The body of knowledge of Software Engineering (SE) began to be integrated since the late sixties in response to a situation known as "the software crisis". This situation generated dissatisfaction in customers mainly for three particular situations linked to the software process: (1) cost overruns in the software process, (2) non-compliance with the delivery deadlines, and (3) requirements for the delivered software product are not met. After just half a century, Software Engineering (SE) has a body of knowledge accepted by professionals and researchers of the discipline [1], which integrates a set of areas linked to development processes (Requirements, Design, Construction, Testing and Maintenance) and management processes (Quality, Configuration, Software Process) associated to the aforementioned areas. There is also a set of methods, techniques, tools and good practices that have been integrated into the body of knowledge of the SE with the intention of improving both the processes as the products generated through said discipline.

© Springer Nature Switzerland AG 2019
J. Mejia et al. (Eds.): CIMPS 2018, AISC 865, pp. 69–79, 2019.
https://doi.org/10.1007/978-3-030-01171-0_6

The improvement of the software process has been studied from several angles, one of which has had singular interest in recent years, studies the importance of the human factor in the software development process; this factor acquires singular relevance due to the fact that several of the processes linked to Software Engineering are carried out in the context of work teams, and therefore, the human factor can affect the success or failure of a project of this kind. In this sense, De Marco comments in [2] that the main problems or causes of the failure of the projects are not of a technological nature, but rather are due to factors of a sociological nature; for his part, Humphrey in [3] states that the process of forming and building a software development team does not happen by accident, the team needs to establish working relationships, agree on objectives and determine roles for group members.

The purpose of the study described in this article is to explore whether the use of role theory proposed by Belbin [4] for the formation of effective teams has any influence on the measurements made by development teams trained with this theory, in comparison with the measurements made by traditional development groups, integrated randomly. It is worth mentioning that the controlled experiment was developed within the framework of a course on Design of Experiments given to students of a Bachelor in Software Engineering.

The remaining part of this document is organized as follows: Sect. 2 presents a general description of related work. Section 3 briefly presents the software measurement technique known as Function Points. Section 4 presents the experiment performed. Section 5 presents the results end findings. Finally, the conclusions are presented in Sect. 6.

## 2 Related Work

Some researchers [4–6] claim to have identified roles that describe the behavior of individuals in work teams—team roles—and although there is no evidence that they are particularly associated with any type of activity, their absence or presence is says that it has significant influence on the work and the achievements of the team [7].

Among the proposals on team roles, Belbin's work [4] is surely the most used among consultants and trainers, its popularity is that, it not only offers a categorization of roles, but also describes a series of recommendations for the integration of work teams, which are known as the Belbin Role Theory. In relation to the roles of Belbin, Johansen in [8] proposed to validate the inventory of self-perception proposed by Belbin through the observation of the behavior of work teams. As part of his conclusions, he indicated that it is worthwhile to use said instrument as a tool to evaluate the composition and possible performance of the team.

Pollock in [9] explores whether diversity of roles and personalities among the members of a team of information systems students can improve the effectiveness of the team. He concludes that diversity does not have a significant influence on the effectiveness of the team; however, he says that the presence of certain roles, such as the Sharper, Chairman and Completer-Finisher, can increase effectiveness. In this same sense, Henry and Stevens [10] reported a controlled experiment with students in which they explore the improvement in the effectiveness of software development teams, based on the set of roles proposed by Belbin; these authors analyzed the general usefulness of Belbin's roles—particularly the roles that the leader can play—in terms of two aspects: performance and viability; they conclude that teams that contain only one leader role present better performance than those that do not include it or those that include more than one leader; in a similar study, Estrada and Peña [11] reported that some roles have a greater contribution with certain activities, as was the case of the Implementer role in the coding task. Aguilar in [12] reports that the collaboration skills presented by groups formed on the basis of role theory (GBF), is significantly greater than that presented by the groups formed solely taking into account functional roles (GMF); reports that GBFs spend more time than GMFs in their work sessions. In a previous work [13] we compared, in an academic environment, the quality of the readability of the code generated by teams integrated with the Belbin Theory (EB), in contrast to randomly-formed teams (ET); among the results obtained, it was reported that EB teams show significantly better results than ET teams.

## 3 Function Point Analysis

The Function Point Analysis (FPA), introduced by Allan Albrech in 1979, is an accepted standard for measuring the logical or functional size of software projects or applications, based on the functional requirements agreed upon with the user [14]. The general process for measuring the size of a software system can be summarized in the following steps:

1. *Determine the type of Function Point count to be conducted:* This technique allows to measure the functionality at the beginning of the development process, in maintenance time or when a system is already in operation time.
2. *Identify the application boundary.* Identification of the scope is established in the start phase of the project, it is usually documented in the form of software requirements or use cases.
3. *Identify all data functions and their complexity.* There are two types of functionality linked with logical files.

   - *Internal Logical File (ILF):* is a group of logically related data or control information maintained through an elementary process of the application within the boundary of the software system.
   - *External interface file (EIF):* is a group of logically related data or control information referenced by the application but maintained within the boundary of different application.

4. *Identify all transactional functions and their complexity.* It is possible to identify three types of transactions:

   - *External input (EI):* is an elementary process of the application which processes data or control information that enters from outside the boundary of the software system.
   - *External Output (EO):* is an elementary process of the application which processes data or control information that exits the boundary of the software system.

- *External Inquiries (EQ):* is an elementary process of the software system which is made up of an input-output combination that results in data retrieval.

5. *Determine the Unadjusted Function Point (UFP) count.* According to IFPUG [15], the complexity of each of the five types of functions can be determined using a complexity table (see Table 1).

**Table 1.** Unadjusted function point counting weights.

| Type | Low | Average | High | Total |
|------|-----|---------|------|-------|
| EI | __ x 3 | __ x 4 | __ x 6 | |
| EO | __ x 4 | __ x 5 | __ x 7 | |
| EQ | __ x 3 | __ x 4 | __ x 6 | |
| ILF | __ x 7 | __ x 10 | __ x 15 | |
| EIF | __ x 5 | __ x 7 | __ x 10 | |

*Total UFP*

6. *Determine the Value Adjustment Factor (VAF).* An adjustment factor is calculated that considers fourteen general aspects related to the system (i.e. reusability, ease of change, range of transactions).

7. *Calculate the final Adjusted Function Point (AFP) count.* VAF is applied to the Unadjusted Function Point count.

## 4   Methodology

A controlled experiment was designed to explore the influence of the use of the Belbin Role Theory in the integration of software development teams. In our case, the influence was evaluated using the result—product—of applying the technique of function points. In particular, the five metrics linked to the functionality of the files (ILF, EIF) and transactions (EI, EO, EQ) declared in an Software Requirements Specification were compared. Additionally, the time used by the teams to carry out the task was used as a dependent variable—a metric process.

### 4.1   Objective, Hypothesis and Variables

With the aim of exploring whether the integrated work teams based on the Belbin Role Theory—which we will call Compatibles—generate measurements significantly different from those obtained by the integrated teams without using any particular criteria —which we will call Traditional—five pairs of hypotheses—a pair for each of the metrics obtained with the function point technique—were proposed, such as the following:

$H0_1$:   The average of the ILF obtained with the Function Point Technique by the Traditional Teams is equal to the average of the ILF obtained by the Belbin Teams.

H1₁:    The average of the EIF obtained with the Function Point Technique by Traditional Teams differs from the average of the EIF obtained by the Belbin Teams

Likewise, with the objective of identifying whether the time spent by the integrated software development teams based on the Belbin Role Theory, in the aforementioned software measurement task, differs from the time that the integrated teams invest randomly, we proposed the sixth hypotheses:

H0₆:    The average time recorded by the traditional teams in the measurement task is equal to that reported by the Belbin Teams.

H1₆:    The average time recorded by traditional teams in the measurement task differs from that reported by the Belbin Teams

The factor to be controlled in this experiment is the integration mechanism of the software development teams, which has two alternatives: (1) Belbin Teams, and (2) Traditional Teams. On the other hand, the six response variables (ILF, EIF, EI, EO, EQ, Time) obtained by applying the technique of Function Points in the ERS, will be recorded in the instrument that the work teams deliver at the end of the activity. Aspects such as the complexity of the problem to be solved, the time available for the task, as well as the degree of experience of the participants, are considered parameters that do not affect or skew the results of the study, because they are homogeneous parameters for all the teams of development; in the case of the degree of experience, teams are composed of students who are in their training process and, at the time of the experiment, still unaware of the technique, are also homogeneous.

## 4.2  Participants/Subjects

The participants in the experiment were twenty seven students of a Bachelor in Software Engineering from the Autonomous University of Yucatan, who were studying the Software Engineering Experiments Design course in the August-December 2016 semester, subject located in the fifth semester.

With the students enrolled in the course, nine software development teams of three members each were formed; we used the information obtained—primary roles in students—after the administration of the self-perception inventory of Belbin, and we integrate five teams with compatible roles (Belbin Teams: BT) and four additional teams with students assigned in a random way (Traditional Teams: TT). Given that the measurements would be obtained on the products generated by the development teams, the experimental subjects in this case were the nine work teams integrated by the researchers. The conformation of the five teams (with three members) based on the Belbin Role Theory are described in Table 2.

## 4.3  Experimental Design

A Factorial Design with "one variation at a time" was used, the independent variable corresponds to the way of integrating the work teams, in the study there are two

**Table 2.** Integrated teams with Belbin Theory (Compatible teams).

| Teams | Integrating roles |
|-------|-------------------|
| I | (1) Implementer, (2) Monitor-Evaluator, (3) Coordinator |
| II, IV | (1) Implementer, (2) Implementer, (3) Completer |
| III, V | (1) Team Worker, (2) Team Worker, (3) Plant |

**Table 3.** Experimental subjects by treatment.

| Treatment | Teams |
|-----------|-------|
| Belbin Teams (BT) | I, II, III, IV, V |
| Traditional Teams (TT) | VI, VII, VIII, IX |

experimental treatment factor levels: BT and TT. Table 3 illustrates the assignment of the nine teams to each of the treatments.

### 4.4   Execution of the Study

The study was carried out in three work sessions; in the first one, the self-perception study was administered to the students, this session was carried out in the last half hour of a class session of the subject. Subsequently, in a second session that lasted two hours, participants received instruction on the technique function point analysis. In the third session, the experiment was executed.

As the first activity in the experimental session, the nine teams were integrated based on what was described in Sect. 5.2, and they were provided with the ERS of a case study, as well as a report sheet with the scheme of Table 2; brief description of the activity was described and they were asked to identify the team, as well as record the start and end time of the activity.

## 5   Analysis and Results

This section presents both the descriptive statistical analysis of the measurements collected and the inferential statistical analysis.

For the descriptive analysis, due to the limitations of space in the publication, six simultaneous boxplot were generated (see Fig. 1); we selected that type of graph because they represent a descriptive way of comparing treatments. In Fig. 1 we can see that only the graphics (a) and (f) do not present overlaps in the treatments, so it is very likely that they present differences between them; also, in the first case, a measure obtained for EO, the BT average was closer to the real value. We can also observe that in the two measurements linked with data functions, BT present greater dispersion and asymmetry than TT; As for the other two measurements related to transactions, no differences were observed in the dispersion of the treatments, however the BT have outliers in both cases.

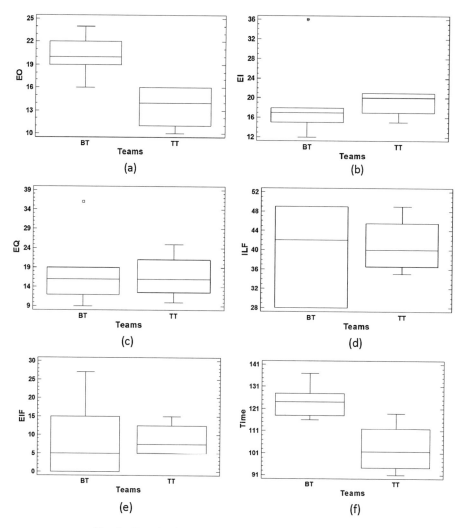

**Fig. 1.** Boxplot for each of the six comparative analyzes.

In order to evaluate the differences observed in the EO variables—as a product metric—and Time—related to the process—with the descriptive analysis and determine if they are significant from the statistical perspective, the following statistical hypotheses were raised:

$$HO_{EO} : \mu_{BT} = \mu_{TT}; H1_{EO} : \mu_{BT} < > \mu_{TT} \tag{1}$$

$$HO_{Time} : \mu_{BT} = \mu_{TT}; H1_{Time} : \mu_{BT} < > \mu_{TT} \tag{2}$$

We proceeded to use the one-way ANOVA; the associated linear statistical model is the following:

$$Y_{ij} = \mu + \beta_i + \varepsilon_{ij};$$

where $Y_{ij}$ is the ij-th observation (value of the j-th replica under treatment i), $\mu$ is a parameter common to all treatments called general or global mean, $\beta_i$ is a parameter associated with the i-th treatment called effect of the i-th treatment and $\varepsilon_{ij}$ is the random component of the error. Tables 4 and 5 present the results of these analyzes.

**Table 4.** ANOVA for EO by teams.

| Source | SS | Df | MS | F | P-Value |
|---|---|---|---|---|---|
| Between groups | 99.7556 | 1 | 99.7556 | 10.94 | 0.0130 |
| Within groups | 63.8 | 7 | 9.11 | | |
| Total (Corr.) | 163.556 | 8 | | | |

**Table 5.** ANOVA for time by teams.

| Source | SS | Df | MS | F | P-Value |
|---|---|---|---|---|---|
| Between groups | 1012.94 | 1 | 1012.94 | 9.96 | 0.0160 |
| Within groups | 711.95 | 7 | 101.707 | | |
| Total (Corr.) | 1724.89 | 8 | | | |

The ANOVA table decomposes the variance of the variable under study into two components: an inter-group component and an in-group component. Since the P-value of the F-test is less than 0.05 in both study variables (EO:0.013 & Time:0.016), we can reject the null hypothesis and affirm in both cases that there is a statistically significant difference between the mean of the variable under study between a level of Treatment and other, with a 5% level of significance.

$$H1_{EO} : \mu_{BT} < > \mu_{TT} \ \& \ H1_{Time} : \mu_{BT} < > \mu_{TT}$$

The ANOVA Model has associated three assumptions that it is necessary to validate before using the information it offers us; the assumptions of the model are: (1) The experimental errors of your data are normally distributed, (2) Equal variances between treatments (Homoscedasticity) and (3) Independence of samples.

To validate the first assumption, we will use the normal probability graph. It is a graphical technique for assessing whether or not a data set is approximately normally distributed. As can be seen in the graph of Fig. 2, the points, in both graphs, do not show deviations from the diagonal, so it is possible to assume that the residuals have a normal distribution in both cases.

In the case of Homoscedasticity, we generate a residuals vs. fitted plot and we observed if it is possible to detect that the size of the residuals increases or decreases systematically as it increases the predicted values; as we can see in the two graphs of

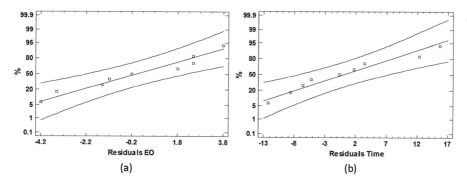

**Fig. 2.** Normal probability plot for (a) EO and (b) Time.

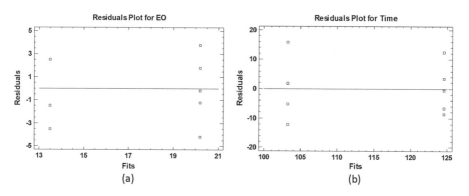

**Fig. 3.** Residuals vs. Fitted plot for (a) EO and (b) Time.

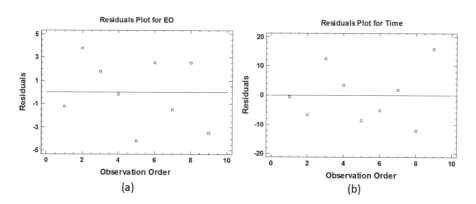

**Fig. 4.** Residuals vs. Order of the data plot for (a) EO and (b) Time.

Fig. 3, no pattern is observed by which we can accept that the constant variance hypothesis of residuals is met.

Finally, to validate the assumption of data independence, we generate a residuals vs. Order of the Data plot; in this case, we observe if it is possible to detect any tendency to have gusts with positive and negative residuals; in the case of our analysis, we can see in Fig. 4 that in both cases no trend is identified, so it is possible to assume that the data come from independent populations.

## 6   Conclusions

In this paper, we present a controlled experiment in which we compare the performance —metrics linked to the product and the process—of work teams in tasks related to software development, particularly, software measurement using the Function Point Technique. The two treatments to be compared, were linked with the way of integrating the work teams; firstly, the traditional way of randomly assigning its members, and on the other, the proposal to integrate teams using the Belbin Role Theory. The results regarding the process metrics showed significant differences only in the EO metric; however, no prior information is available that explains in any way the observed result; that is, it is not clear that the type of activity carried out for the measurement of the EO was particularly different than that used in the other metrics, such that this could be influenced by the type of conformation of the team; what could be observed, is that although the average value in both types of teams was far from the real value, the BT averaged a value closer to the real, which coincides with that reported in [13]. Regarding the metric associated with the process, in this case, the time that the teams used to carry out the task, the results showed significant differences; that is, the BT use more time than the TT, which coincides with one of the experiments reported in [12], in the sense that the teams formed according to the Belbin Role Theory develop in their work sessions a greater degree of interaction, which means that they invest more time in trying to reach a consensus in the decisions of the team.

## References

1. Bourque, P., Fairley, R.: Guide to The Software Engineering Body of Knowledge (SWEBOK V3.0). IEEE Computer Society (2014)
2. DeMarco, Y., Lister, T.: Peopleware Productive Projects and Teams, 2nd edn. Dorset House Publishing Co., New York (1999)
3. Humphrey, W.: Introduction to the Team Software Procesess. Addison Wesley Longman Inc., Reading (2000)
4. Belbin, M.: Team Roles at Work. Elsevier Butterworth Heinemann, Oxford (1993)
5. Mumma, F.S.: Team-Work & Team-Roles: What Makes Your Team Tick?. HRDQ, King of Prussia (2005)
6. Margerison, C.J., McCann, D.J.: Team management profiles: their use in managerial development. J. Manag. Dev. 4(2), 34–37 (1985)
7. Senior, B.: Team roles and team performance: is there 'really' a link? J. Occup. Organ. Psychol. 70, 85–94 (1997)
8. Johansen, T.: Predicting a Team's Behaviour by Using Belbin's Team Role Self Perception Inventory. Dissertation at Department of Management & Organisation. University of Stirling (2003)

9. Pollock, M.: Investigating the relationship between team role diversity and team performance in information systems teams. J. Inf. Technol. Manag. **20**(1), 42–55 (2009)
10. Henry, S., Stevens, K.: Using Belbin's leadership role to improve team effectiveness. J. Syst. Softw. **44**(3), 241–250 (1999)
11. Estrada, E., Peña, A.: Influencia de los roles de equipo en las actividades del desarrollador de software. ReCIBE **2**(1), 1–19 (2013)
12. Aguilar, R.: Una Estrategia Asistida por Entornos Virtuales Inteligentes. Doctoral Thesis. Polytechnic University of Madrid (2008)
13. Aguileta, A. Ucán, J., Aguilar, R.: Explorando la influencia de los roles de Belbin en la calidad del código generado por estudiantes en un curso de ingeniería de software. Revista Educación en Ingeniería **12**(23), 93–100 (2017)
14. Garmus, A., Herron, D.: Measuring the Software Process. Prentice Hall, Upper Saddle River (1996)
15. IFPUG: Function Point Counting Practices Manual Release 4.1.1. International Function Point Users Group (2010)

# Proposal of a Model for the Development of Labor Competencies Based on Serious Games in the Context of Industry 4.0

Madeleine Contreras[1]([⊠]), David Bonilla[2],
and Adriana Peña Pérez Negrón[2]

[1] Universidad de Guadalajara, Sistema de Universidad Virtual,
Av. Enrique Díaz de León no. 782, Col. Moderna,
CP 44190 Guadalajara, Jalisco, Mexico
mgabriela@suv.udg.mx
[2] Universidad de Guadalajara, CUCEI,
Blvd. Marcelino García Barragán #1421, esq Calzada Olímpica,
CP 44430 Guadalajara, Jalisco, Mexico
jose.bcarranza@academicos.udg.mx,
adriana.pena@cucei.udg.mx

**Abstract.** Nowadays, the importance of video games in education is undeniable. The benefits of using video games for classroom instruction have been proven in many studies. Active learning methods promote the development of skills that simulate real-life situations and problems, help managing costs more effectively and generate interest and involvement in employees. However, these benefits have not been broadly explored in some domains, as is the industry case. In this paper is presented a learning model based on serious games, thought-out to industrial production plants, like an a viable alternative for the development of labor competencies.

**Keywords:** Labor competencies · Serious gaming · Industry 4.0

## 1 Introduction

In the past, the three industrial revolutions took place due to technical innovations. At the end of the 18th century, the introduction of mechanical manufacture driven by water and steam, the division of labor at the beginning of the 20th century, and the introduction of programmable controller logic (PLC) for automation as a manufacturing technique in the 70s. According to the industry and the academy experts, the next industrial revolution is being triggered by the Internet, which allows communication between humans and machines in Cyber-Physical-Systems (CPS) through large networks [1].

Industry 4.0 represents the special interest in production processes and processes that involve the so-called intelligent manufacturing, which on the one hand supports products development in a faster and flexible way. And on the other hand, the concept of Industry 4.0 recognizes the complexity of the new industry, integrating factors such as automation in those processes and its personalization.

© Springer Nature Switzerland AG 2019
J. Mejia et al. (Eds.): CIMPS 2018, AISC 865, pp. 80–87, 2019.
https://doi.org/10.1007/978-3-030-01171-0_7

Innovation can be announced as a revolution when it promotes changes and transformations in different areas activity. Accordingly, human activity is constituted in the material, the external, in the actual management of objects, which in turn is based on the significant of the learning process, closely linked to the component of cognitive representation [2]. Therefore, rethinking the development of skills in a work environment, as it is the industry, involves recognizing the imminent genesis towards proposals that would break the mechanistic paradigms of the industry beginnings.

In this context, the use of alternatives for the industry like incorporating Virtual Reality (VR) seems like a natural step. Brudniy and Demilhanova [3] defined VR as the most advanced form of relationship between a person and a computer system. A relationship that allows a direct interaction between the user and the computer generated environment.

The continuous technology development and its strong penetration in the society in general, and the video games in particular, are unquestionable. In recent years, the use of video games for educational purposes is a field of interest that has been constantly growing. The concept of video games can be applied for teaching in two forms: gamification and serious games.

Gamification consists in applying the principles of the design of the video games, the use of the mechanics and the own elements of a game in any process, beyond the own context of the video games. The goal is to take advantage of both the psychological predisposition of people to participate in games and the very benefits of the game to motivate and improve or change the behavior of the participants.

On the other hand, serious games are video games, but unlike the latter, serious games are developed under a specific intention, which can be a real problem simulation or transmitting a message, both seeking to reach a specific group of players, serious gaming has other than ludic purposes [4].

The concept of serious gaming tends to the primary logic of the game, and thus its playful nature is also incorporated. Huizinga [5] defined it as a free activity that is consciously kept out of ordinary life for lack of seriousness, but at the same time deeply absorbing for the person, who exercises it. Serious games offer an atmosphere constituted by learning environments in which to experience real problems.

Serious games represent an advantage for the development of competences because they allow practicing in a safe environment while playing, rehearsing in situations based on real life. The motivation that encourages the playful action itself has an important effect on learning. In addition to the learning of specific tasks or training, this type of games also supports the acquisition of certain skills or competencies. Additionally, the use of just gamification might not provide for the user a total involvement in the game when compared to serious games.

Next, the proposal to integrate serious games under a defined theoretical perspective will be explained. The ecosystemic with a concrete model that goes through several stages from the design, implementation and evaluation of the method is explained. It is combined with a competency-based approach focused in the capacity of action achieved by the learner, through the mobilization of resources, such as knowledge, skills, techniques and behaviors [6].

## 2  Proposal of the Inclusion of Labor Competencies Using Serious Gaming in Industries 4.0

In order to present a general model, we intend to offer a vision that promotes the incorporation of serious gaming in industries 4.0. Recognizing what Prenski [7] visualized, the omnipresence of digital technology in today's society, serious gaming represents the possibility of integrating the pedagogical and the technological; encompassing processes of training and staff training, with respect to diversity, and from an anchored perspective resulting from virtual learning environments. In order to proportionate independent, autonomous and self-regulated learning [8].

In Fig. 1 is shown the possibility, offered by serious games, as a modality for training, where participants constantly get feedback on this experience. From the pedagogical point of view, the labor competencies are adapted to the needs of the organization; they are certainly valued as an expected product, despite the inclusion of the technological component. Offering the opportunity to innovate and achieve transformations that lead to advancement in said competences.

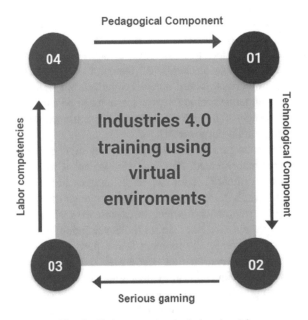

**Fig. 1.** Serious gaming in Industries 4.0

Schuldt and Friedemann [9] proposed the integration of gamification to the Industry 4.0.; using the Günthner et al. [10] process, they determined its integration stages. For Schuldt and Friedemann is relevant to take care, when implementing serious gaming, that the usual flow of work is maintained. It should not be substantially interrupted with distractions such as sounds or animations; those should be eliminated as much as possible. Another important condition is handling data confidentiality, both in the design and in the information derived from the participants. Therefore, confidentiality policies must be declared to protect both the system and the users. Last but not least, they recommend a selective use of serious gaming to avoid a behavioral habituation to them.

The Schuldt and Friedemann proposal for the integration of gamification is here taken as a base and adapted for serious games and competencies development, see Fig. 2. Their proposal was modified in the description, selection and development section; elements that involve the development of specific skills to objectives achievement in the incorporation of competencies development in the gamification process.

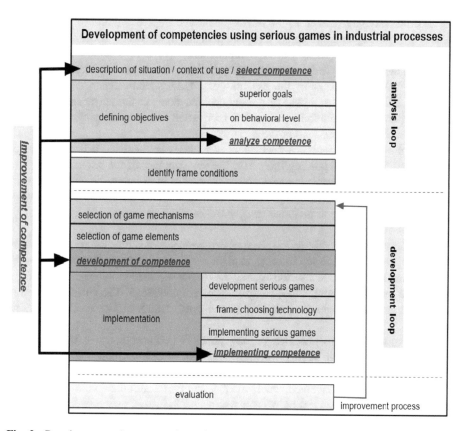

**Fig. 2.** Development of competencies using serious games in industrial processes, modified from Schuldt and Friedemann [9].

Taking into consideration the ecosystemic approach as theoretical support, this paper purpose is not to deepen in it, but to state a theoretical position. As stated in two classics, Humberto Maturana and Francisco Varela [11], they sustain that the systems of living beings (people), are always coupled to their environment, aimed to the preservation of their organization. Therefore, it was conceived in this first stage of the design, a learning situation, with problems and objectives that are inserted in the specific work environments. The interactions from people directed towards the activity, the devices used, and the processes adaptive, among others.

In the first instance, as can be observed in Fig. 2, competencies are based on objectives that must be clearly defined, and well known at any time by the participant player in the serious game. These are the learning objectives of the competencies that are required at work.

Also in the first stage of the model, which can be identified in Fig. 2 as an analysis loop, it is emphasized the objectives that in the literature on gamification are related as challenges, that the proposal suggests be adopted as part of the jargon of the serious game itself [12].

For the next stage linked to the development and implementation of the competition, it is necessary to contemplate three fundamental components of the serious game: the dynamics, the mechanics and the components.

When the dynamics are described, we start with some previously identifiable ones. They are linked to aspects of the psychological field, specifically to the anticipated emotions. It is possible to arouse or present spontaneously in the player, as well as the attitudes, perceived limitations or manifest, motivation linked to the progression of the challenges; narratives are among the most relevant.

Regarding the mechanics, aspects such as rewards or feedback take place, which can be identified as the guiding thread for the development of the game.

Lastly, the third element important to be integrated in this stage of development, is precisely that identified as components, which arise from the relationship between dynamics and mechanics, included in this section such as avatars, badges, rankings, levels, and the teams.

In order to generate motivation, *badges* can be incorporated as the recognitions for the participants that stand out in some modality of the expected performance, thus the *ranking lists* also work. In a serious game, the player will face problems that will force him/her to look for solutions that contribute solve a challenge. The closing criteria of the partial and final stages are proposed (game over), which will be result in learning outcomes.

The proposal aim to integrate serious games for competence development, as a learning alternative that begins with the selection of one or more skills, which as mentioned in the previous section, [13] pay special attention to the action. In addition, competition is also conceived, such as the ability to mobilize various cognitive resources, skills and attitudes as we can see in Fig. 3.

The Schuldt and Friedemann [9] process of gamification integration includes at first level the description of the situation where the employee is expected to deploy a specific competition. Therefore, in this first stage for analysis, the concept of competency should be included, the selection of competencies. In addition to detailing actions, also enunciate skills and attitudes related to a serious game. It is possible to establish in advance the type of attitude that is sought; hence the importance of approaching the ideals, mission, and vision of the industry, implies a stage in the broad sense of analysis [14].

After that, technical aspects of the game that contribute to the achievement of objectives, the structure, elements and intentionally, as well as clear work of the

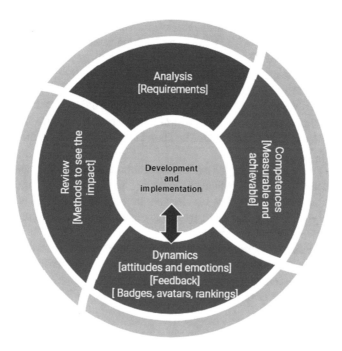

**Fig. 3.** Fundamental components of the serious games in the Development and implementation stage.

competition are detailed. In a final stage, the evaluation of the learning in which they participate in the serious game is contemplated. It has to be considered that the same game already includes this tool and in turn the team dedicated to training-learning can analyze and evaluate the results. This final stage is equally valuable, as it will allow a process of continuous improvement [14].

Adaptation to the Schuldt and Friedemann [9] proposal can be located within the analysis and development cycle, where as a first step is the selection of competencies to be acquired. Once this point has been defined, it is necessary to analyze if the competencies are in concordance with the planned objectives at the moment of describing the context. As a third stage after the selection of game elements, the development of the competencies within the game should be started, such as learning. Finally, the implementation of competencies should have mechanisms by which they can be evaluated.

For this stage of evaluation that also acquires a fundamental importance, the incorporation of methods and techniques of quantitative and qualitative order, such as individual interviews, focus groups, pre-intervention and post-intervention through serious games are planned.

The idea is how the information derived from a clear and precise identification of labor competencies was described from a triangular, starting with the performances achieved in the process of implementing the game in search of improving the quality standards of the organization.

The observation of the performance of the individual in different performed functions in the workplace or in the most similar conditions is possible [15]. This stage of evaluation should include techniques, as above mentioned, that can gather information about knowledge, skills, attitudes and values from the participant, and how he/she applies them in his/job function.

Also, results obtained in the case of production in the industry or the different areas that make up the organization, can be taken into consideration for the evaluation; previous learning in relation to the required competencies, as well as the academic and certifications achievements.

Activities that can integrate this evaluation section are the third parties' reports such as the observation of their immediate superiors or pairs evaluation. It has also been the case, that the same community has benchmarks of the performance of employees who at the time might be fully identified by the organization such as contributions, criticisms or statements in the community.

Finally, the model in congruence with its theoretical ecosystemic and competency-based approach, is about verifying and validating what the worker [6] that participated in this learning experience, through serious games, achieves in terms of resolution of concrete situations, not in terms of knowledge that later can be forgotten or that will lack of work practice.

## 3  Conclusions and Future Work

The industries, institutions and business corporations should realized of the potential of video games. They represent saving time and money, very important issues in organizations. Also, they are motivational an effective for training. The serious gaming privileges the position of the participants as active agents in their training, they integrate into one environment that generates more dynamism, weakens the resistance or rejection of users.

In general, researchers as Markopolus, [14] concluded that gamification has a positive effect on learning, making the subjects or, in this case, competencies, which are usually more difficult to develop, softened by the direct effect of intrinsic motivation. It can also be observed favorable results in collaboration, interest, reduction and more optimal management of the workload.

Riedel and Hauge [16] concluded their study by showing how a large part of the gamification offers specific targets achievement, transmitting hard skills such as understanding how complex systems work in commercial and production, as well as interpersonal or soft skills such as collaboration, creativity and social communication, tackling problems that are not usually addressed by other learning in virtual platforms.

It is possible to glimpse an interesting vein in the field of learning in the industry. This proposal could be applied to any company worldwide adapting it to the conditions of each culture and his industry 4.0.

# References

1. Brettel, M., Friederichsen, N., Keller, M., Rosenberg, M.: How virtualization, decentralization and network building change the manufacturing landscape: an industry 4.0 perspective. World Acad. Sci. Eng. Technol. Int. J. Inf. Commun. Eng. **8**(1), 37–44 (2014)
2. Leóntiev, A.N.: Teoría psicológica de la actividad. En A.N. Leóntiev, Selección de Obras de Psicología, Tomo II. Moscú: Pedagogía (en ruso). Moscú (1983)
3. Brudniy, A., Demilhanova, A.: The virtual reality in a context of the "Mirror Stage". Int. J. Adv. Psychol. **1**(1), 6–9 (2012)
4. Darejeh, A., Salim, S.S.: Gamification solutions to enhance software user engagement—a systematic review. Int. J. Hum. Comput. Interact. **32**(8), 613–642 (2016). https://doi.org/10.1080/10447318.2016.1183330
5. Huizinga, J.: Homo Ludens. Alianza Editorial, Madrid (2000)
6. Haddouchane, Z.A., Bakkali, S., Ajana, S., Gassemi, K.: The application of the competency-based approach to assess the training and employment adequacy problem. arXiv preprint arXiv:1704.04985 (2017)
7. Prenski, M.: Digital Natives, Aprendizaje para el nuevo milenio. Universidad Camilo José Cela (2011)
8. Henríquez, G.: Modelo de capacitación docente para entornos virtuales de aprendizaje. Caso decanato ciencias de la salud de la UCLA. Revista Iberoamericana de Educación a Distancia. Madrid (2015)
9. Focusing on man in future machine-driven working environments. In: IEEE Global Engineering Education Conference, EDUCON, pp. 1622–1630 (2017). http://doi.org/10.1109/EDUCON.2017.7943066
10. Günthner, W.A., Mandl, H., Klevers, M., Sailer, M.: GameLog – Gamification in der Intralogistik, 85–89 (2015)
11. Becerra, G.: De la autopoiesis a la objetividad. La epistemología de Maturana en los debates constructivistas. Opción **32**(80), 66–87 (2016). Universidad del Zulia Maracaibo, Venezuela
12. Zepeda-Hernández, S., Abascal-Mena, R., López-Ornelas, E.: Integración de Gamificación y Aprendizaje Activo En El Aula **12**(6), 315–325 (2016). Universidad Autónoma Indígena de México, El Fuerte, México julio-diciembre
13. Solano-Albajes, L. Contreras-Espinosa, R.S., Eguia Gómez, J.L.: VIDEOJUEGOS: conceptos, historia y su potencial como herramientas para la educación. Revista Revista de investigación Editada por Área de Innovación y Desarrollo, S.L (2013)
14. Markopoulos, P., Fragkou, A., et al.: Gamification in engineering education and professionaltraining. Int. J. Mech. Eng. Educ. **43**(2), 118–131 (2015)
15. Salas Perea, R.S., Diaz Hernandez, L., Perez Hoz, G.: Evaluación y certificación de las competencias laborales en el Sistema Nacional de Salud en Cuba. Educ. Med. Super [online]. **28**(1), 50–64 (2014). ISSN 0864-2141
16. Riedel, J., Hauge, J.: State of the art of serious games for business and industry. In: 17th International Conference on Concurrent Enterprising (2011)

# Reinforcing Very Small Entities Using Agile Methodologies with the ISO/IEC 29110

Mirna Muñoz[1(✉)], Jezreel Mejia[1], and Claude Y. Laporte[2]

[1] Unidad Zacatecas, Centro de Investigación en Matemáticas A.C., Parque Quauntum, Ciudad del Conocimiento, Avenida Lassec, Andador Galileo Galilei, Manzana, 3 Lote 7, CP 98160 Zacatecas, Zac, Mexico
{mirna.munoz,jmejia}@cimat.mx
[2] Department of Software and IT Engineering, École de technologie supérieure, 1100, Notre-Dame Street West, Montréal, Quebec H3C 1K3, Canada
Claude.Laporte@etsmtl.ca

**Abstract.** Software development organizations have had significant economic impact activity in recent years. The growing of the software demand has created opportunities for very small entities (VSEs) to produce software and services to satisfy market needs. This situation highlights the increasing need for improving their development software processes to stay in the market and to develop quality software products and services to achieve steady grown. Unfortunately, a common issue that most of the VSEs are facing is the lack of the knowledge and practical experience regarding the implementation of current software process improvements (SPI) models and standards. This problem becomes critical because VSEs are under increasing pressure to improve their productivity and quality while keeping costs to a minimum. This paper describes experience to reinforce the software development process of 4 software development organizations had used agile methodologies and implemented the ISO/IEC 29110 standard in the state of Zacatecas of México. The paper presents this experience, the benefits, lessons learned and the goals achieved.

**Keywords:** ISO/IEC 29110 · Software industry · Very small entities
VSEs · Project management · Implementation process · Standard
Agile methodology · Scrum

## 1 Introduction

Nowadays, a large percentage of small and medium enterprises (SMEs) are using agile methodologies in an effort to produce software that meets the time requested by the market [1]. However, in most cases the lack of knowledge about how to correctly apply agile methodologies and software engineering proven practices might be contributing to inefficient software development (e.g. in quality, cost and time), and very small entities (VSEs) are not the exception.

Therefore, implementing proven practices contained in software process improvement (SPI) models and standards in real environments of software development organizations represent a big challenge, especially in VSEs (i.e. enterprises, organizations, departments or projects having up to 25 people) which must work harder

© Springer Nature Switzerland AG 2019
J. Mejia et al. (Eds.): CIMPS 2018, AISC 865, pp. 88–98, 2019.
https://doi.org/10.1007/978-3-030-01171-0_8

in order to survive, and must also spend time and effort in improving their operation and processes [2, 3].

This highlights the importance of developing quality products provided by this type of organizations. According to [2] VSEs are developing and/or maintaining the software used for most organizations.

Even when standard organizations such as ISO and IEC provide solutions to help VSEs to implement proven practices such as the ISO/IEC 29110 standard series [4], a common problem that most of them are facing is the lack of knowledge and practical experience regarding the implementation of an SPI model or a standard.

This paper aims to describe the experience of a group of VSEs that improved their productivity and quality while keeping costs to a minimum by using the ISO/IEC 29110. In México, this standard has been adopted as one of the Quality Standards that have the recognition of the government and the industry [5].

After the introduction, this paper is structured as follows: Sect. 2 shows an introduction of agile methodologies; Sect. 3 shows an introduction of ISO/IEC 29110; Sect. 4 shows related works regarding the implementation of ISO/IEC 29110 in other countries; Sect. 5 describes our five-step improvement method and the experience of implementing the ISO/IEC 29110 in a group of 4 software development VSEs of Zacatecas, México; and in Sect. 6 we present a discussion, conclusions and the next steps.

## 2 Agile Methodologies

Known as lightweight methods, the agile methods are characterized by short, iterative development cycles, performed by self-organizing teams, which use techniques such as simpler designs, code refactoring, test-development, frequent customer involvement. The agile method emphasizes on providing a demonstrable working product with each development cycle [6].

According to the agile alliance, Agile is the ability to create and respond to changes in order to have succeed in uncertain and turbulent environment [7].

Agile software development aims to develop software faster, incrementally and producing satisfied customer. To achieve these objectives, agile methods provide a conceptual framework of practices and principles [7]. The agile methodologies define how to develop software under agile values and principles of the agile manifesto such as iterative development, frequent and early delivery of working software and simplicity [1, 6, 7].

These methodologies emphasized a close collaboration between the development team and the business stakeholder, frequent delivery and self-organizing teams.

There are many agile methodologies, some of the most popular are Rapid Application Development (RAD); eXtreme Programming (XP), Scrum and Feature-Driven Development (FDD) [8].

# 3   ISO/IEC 29110 Systems and Software Engineering Series

As one solution to the challenges of VSEs regarding pressure to improve their productivity and quality while keeping costs to a minimum, the ISO working group 24 (WG24) developed the ISO/IEC 29110 series [5].

The ISO/IEC 29110 series have been designed to help VSEs that develop systems or software. In the context of ISO/IEC 29110, systems are typically composed of hardware and software components. The ISO/IEC 29110 series of software standards and management and engineering guides have been developed to help VSEs to improve their software development process, helping them in the implementation of proven practices focused in the VSEs, in order to get benefits such as increasing their product quality, reducing their delivering time, and reducing their costs of production.

The ISO/IEC 29110 series for software has the following features:

- It contains a set of 4 profiles to be used by VSEs according to their goals: The Generic Profile Group is a four-stage roadmap, called profiles: VSEs targeted by the Entry Profile are VSEs working on small projects (e.g. at most six per-son-months effort) and start-ups. The Basic Profile targets VSEs developing a single application by a single work team. The Intermediate Profile is targeted at VSEs developing more than one project in parallel with more than one work team. The Advanced Profile is target to VSEs that want to sustain and grow as an independent competitive system and/or software development business. Nowadays, the Basic profile it's the only profile in which a VSE can be certified.
- It provides, as a foundation, two processes, the project management process and the software implementation process.
- It can be used to establish processes in VSEs using any development approach, methodology or tool.
- It provides a set of process elements such as objective, activities, task, roles and work products.

The software engineering basic profile is composed by two processes [4, 5, 9]: the project management process and the software implementation process. Each process is composed of a set of activities and each activity is composed of a set of tasks. Table 1 describes the processes, their purpose and their activities

**Table 1.**  ISO/IEC 29110 processes of the software engineering basic profile

| Process | Purpose | Activities |
|---|---|---|
| Project management | Establishes and carries out the tasks related to a project implementation in a systematic way, so that the project's objectives are complying with the expected quality, time and costs | Project Planning<br>Project Plan execution<br>Project Assessment and Control<br>Project Closure |

*(continued)*

**Table 1.** (*continued*)

| Process | Purpose | Activities |
|---|---|---|
| Software implementation | Performances in a systematic way the activities related to the analysis, design, construction, integration and test according to the requirements specified of new or modified software products | Software implementation Initiation<br>Software requirement analysis<br>Software component identification<br>Software construction<br>Software integration and test<br>Software delivery |

## 4 Related Works

Many authors have published success cases in the implementation of ISO/IEC 29110 since it has been adopted in many countries. Next, a set of success cases are briefly described.

In [10] there were reported seven success cases of the implementation of ISO/IEC 29110 in different countries: one from IT start-up from Peru and six from Canadian (2 IT start-ups, one of them with location in Canada and Tunisia; 1 large Canadian financial institution; 1 from automotive domain, 1 from transportation and 1 division of a large American engineering company).

In [11] there were reported pilot projects one from Canada conducted with an IT department with a staff of 4; one from Belgium in a VSE of 25 people; one in France that builds and sells counting systems about the frequenting of natural spaces and public sites with a 14-people VSE. It also mentions some projects executed by graduates and undergraduates' students of the ÉTS (École de technologie supérieure), one in an engineering firm having over 400 employees and other one from a website developed by a VSE of 2 people.

In [12] there were reported the implementation of ISO/IEC 29110 in two very small software development companies in Perú.

In [13] a previous work of the authors of this paper, it reported the implementation of ISO/IEC 29110 in eleven VSEs: one IT start-up; An IT start-up with locations in Canada and Tunisia, A development team at a Canadian IT start-up; four VSEs in México; one software developers at a power train manufacturer; one project teams in a large engineering company's Transmission & Distribution of Electricity division; one software team at a large public utility and a software team in a large financial institution's IT division.

All of them have demonstrated very good results in the implementation of the ISO/IEC 29110, such as reduce rework, access to new customers, increase quality, improve their processes among others. However, even if México has a high number of VSEs certified, i.e. having 31 of a total of 38 VSEs certified in ISO/IEC 29110 [14], there are very few case studies published from México which is the target of this paper.

## 5   Experience in Reinforcing VSEs by Implementing the ISO/IEC 29110

In this section, we describe the experience of implementing the ISO/IEC 29110 in a group of 4 software development companies of Zacatecas that are using the Scrum methodology and are certified by auditors of NYCE, the Mexican Certification Body

### 5.1   Problems Identified in the Four VSEs

An analysis of the VSEs was performed to identify gaps they have in their development processes. It is important to mention that the four organizations use Scrum as base methodology to perform their projects. The problems identified were classified according to the ISO/IEC 29110 processes as follows:

(a) Problems related to project management process: this process covers the planning, execution, assessment and control, and closure of a project. Table 2 shows the detected problems regarding each process activity.

**Table 2.** Problems classified by activity of the project management process

| Process | Problems |
|---|---|
| Project planning | • In most of the cases, they do not have a formal way to receive a request of work by the customer, they do it just by talking with their customer<br>• They do not develop a project plan including resources and training needs |
| Project plan execution | • They do not have an evidence of the reviews performed to know the project progress<br>• Most of them do not have information about plan versus actual values |
| Project assessment and control | • They do not have a formal way to register a change request<br>• They do not register agreements of meetings in a formal way as well as the performed meetings<br>• They do not have a control/track of corrective actions applied in case of significant deviations<br>• They have a lack in the use of baselines<br>• They do not perform a configuration management or it depends of the features of the software tools they are using<br>• They do not perform software quality assurance |
| Project Closure | • They do not have a formal way to get the customer approval |

(b) Problems related to software implementation process: this process covers the software implementation initiation, software requirement analysis, software component identification, software construction, software integration and test, and software delivery. Table 3 shows the detected problems regarding each process activity.

**Table 3.** Problems classified by activity of the software implementation process

| Process | Problems |
|---|---|
| Software implementation initiation | • They do not have evidence the project plan review with team members through which common understanding and commitment to the project is achieved |
| Software requirements analysis | • They do not have an evidence of changes to the requirements |
| Software component identification | • In most of the cases, they do not perform an architectural design as well as a software design |
| Software construction | • They do not document the unit tests performed to the software components |
| Software integration and test | • They do not have evidence of the test procedures followed<br>• They do not register test results<br>• They do not produce verification and validation results |
| Product delivery | • Lack of evidence of the product delivery and acceptation |

## 5.2 Method Used to Reinforce the VSEs with the Implementation of the ISO/IEC 29110

To be able to reinforce the four VSEs, we performed a five-step method:

1. *Identify the main problems of VSEs have performing their projects as they used to work*: this activity is related to having a meeting in which the VSE talked about the way the perform its projects was performed. We addressed this meeting helping us with a questionnaire which contains questions related to project management and software development.
2. *Map the actual processes of VSEs to the ISO/IEC 29110 Basic profile processes*: this activity is related to ask the VSE to identify the way they perform their project management and software development. To achieve it, a sheet with the processes, activities and tasks provided in the ISO/IEC 29110 were provided to the VSE, and then in a second meeting were performed.
3. *Identify and formalize the VSEs proved practices*: this activity is related to provide support to the VSE, so that they identify the practices they are performing regarding the project management and software implementation according to those practices provided in the ISO/IEC 29110 standard. It helps to make aware them on those practices that are still performing within their organization. To show the information a third meeting were performed.
4. *Select and adapt the practices provided by the ISO/IEC 29110 standard to the context of each VSE*: this activity refers to identify the gaps between their actual practices and those provided in the ISO/IEC 29110 standard. Then, once their practices were identified and formalized, a fourth meeting was performed in which were analyzed set of practices provided by the standard, so that they can understand the importance and impact of its adaptation, besides we provide support to help them to identify how they can tailor the practice to their VSE environment.

5. *Review the projects in which the ISO/IEC 29110 practices were implemented and report the non-conformities with respect to the standard*: this activity refers to review a complete project selected by the VSE. To achieve this activity two meetings were performed. On the one way, during the first meeting the VSE showed the project and we took notes. On the other hand, during the second meeting, we provided a report with the non-conformities detected and after that they showed evidence of some to cover the non-conformities. The non-conformities that could not be resolved, were worked by the VSE to solve them.

It is important to mention that the execution of the five-step method was conducted by a series of about 6 meetings with each VSE with a duration average of 4 hours per meeting.

To perform the meeting there were integrated a team of 6 people from a research center as follows:

- Two people with high experience in process definition and improvement in multimodel environments. They have high experience in managing models and standards such as CMMI, ISO 15504, ISO 12207, Moprosoft and ISO/IEC 29110; other frameworks such as PMP and SWEBOK and the methodologies such as TSP, Scrum, XP and crystal.
- Four people with knowledge in software engineering practices, software tools, the CMMI model and the ISO/IEC 29110 standard and in the Scrum methodology.

Besides, each VSE integrate a team according to its characteristics as Table 4 shows.

Once the VSE start the certification process three more meetings were performed but together with the four VSE, the first after the branch analysis, the second after the pre-auditory and the last one after the auditory.

### 5.3   The Four VSEs Reinforced by the Implementation of ISO/IEC 29110

This section presents the four VSEs that were reinforced by the ISO/IEC 29110. It is important to mention that an overview of this 4 VSEs was previously included in [13]. However, this paper analyzes in detail the information regarding the benefits detected in the VSEs.

It is important to highlight that the 4 VSEs achieved the certification, by independent auditors of NYCE, to the Basic profile of ISO/IEC 29110 standard, Table 5 describes the VSEs, the organization description, the project that was used for certification and the benefits they have identified following the implementation of ISO/IEC 29110.

**Table 4.** VSEs of Zacatecas certified to the basic profile of ISO/IEC 29110

| VSE_ID | Organization' description | People involved in the improvement process |
|---|---|---|
| VSE1 | • Develops hardware and software solutions<br>• Has 12 employees<br>• Has highly-trained specialist in electronic development, and in hardware and software development | 1 person with experience in CMMI-Dev certification in maturity level 2 and he have a Scrum Master certification |
| VSE2 | • Started its operation in 2014<br>• Has 7 employees<br>• Uses technologies and platforms oriented to web and mobile applications<br>• Has its own software products that are used by its customers | 2 people with knowledge in software process. Both have a Scrum master certification |
| VSE3 | • Provides IT services to other organizations<br>• Has 3 employees | 1 person with knowledge in CMMI-Dev and Scrum methodology |
| VSE4 | • Offers hardware, firmware and software solutions<br>• Has 4 employees<br>• Provides different quality products to achieve needs of different productive sectors such as mining, pyrotechnic, educational and technological | 2 people with knowledge in agile practices |

**Table 5.** VSEs of Zacatecas certified to the basic profile of ISO/IEC 29110

| VSE_ID | Project description | Detected benefits |
|---|---|---|
| VSE1 | • System to control the access for the company offices | • Provides a software development cycle with defined steps, work products and roles<br>• Reinforces the project plan information<br>• Reinforces the monitoring and control activities to have the project under control<br>• Enables the control of software and tools to be used in a project<br>• Improves the project versioning and the software delivery |
| VSE2 | • Web system that quotes and compares car insurances | • Improves the software development process<br>• Improves the communication with the customer<br>• Improves the change request<br>• Provides documented forms to be implemented as part of the development cycle<br>• Improves the procedure related to the verification |

*(continued)*

**Table 5.** (*continued*)

| VSE_ID | Project description | Detected benefits |
|--------|---------------------|-------------------|
|        |                     | • Improves the procedures related to validation and approval of documents<br>• Improves the documentation of test |
| VSE3   | • Software to manage medical consultation | • Improves the activities related to project monitoring and control<br>• Improves customer communication by using delivery instructions<br>• Improves the management of risks |
| VSE4   | • Redesign of control systems of permanent magnet engines | • Achieves the implementation of a standardized methodology for project management, so that it is possible to reduce cost (mainly unforeseen costs), as well as the reduction of estimated delivery times<br>• Improves the quality of estimated projects<br>• Enable to place the products more quickly and efficiently way in the market |

## 6 Discussion, Conclusions and Next Steps

Nowadays, most of the VSE uses agile methodologies in an effort to produce software that meets the time requested by the market. However, a lack of knowledge of how to correctly apply agile methodologies might be contributing to inefficient software development. Some of the main problems that a VSE faces when using an agile methodology in both the project management and software implementation was listed in Tables 2 and 3 of the Sect. 5.1.

This highlights the increasing need related to the implementation and use of software engineering practices as key aspect to help software development organizations in improving their performance.

In this contexts and in an effort of providing a solution to this need, the software engineering basic profile of the ISO/IEC 29110 series provides a standard and a set of management and engineering guides that are well accepted by software industry because they provide a set of minimum software engineering practices through two key processes: the project management process and the software implementation process. By the way, in México this standard has have a well acceptation.

This paper presented the experience of reinforcing 4 VSEs using the Scrum methodology with the implementation of the basic profile of ISO/IEC 29110.

The main lessons learned identified from the experience are:

- Start by helping organizations to "identify" and "formalize" what are they doing regarding the project management and software implementation processes, by this way it is possible on the one way, to make aware the organizations of the importance of the processes use. On the other way to help them to adapt the standard to the organizations needs and not to adapt the organization to the standard.

- Provide support to VSEs from the beginning of the implementation until they get the certification, by this way it is possible to understand the needs and doubts regarding the implementation and adoption of software engineering practices within their real environment.
- Maintain feedback meetings throughout all steps. These meetings were per-formed during the performance of the five-steps methods, this allows us to be part of the VSE's team, so that the resistance to change was reduced.
- Provide the technical support available for the VSEs whenever they have questions or doubts, even when, there are a set of guidelines developed to help in the implementation of the ISO/IEC29110, most of VSE feels more comfortable if someone is available to help them.

A few improvement opportunities have also been identified:

- Provide training focused on understanding the ISO/IEC 29110 terminology. One of the main problems is the lack of understanding of the concepts hanging in the standard such as baseline, traceability matrix, verification and validation activities. So that, most of the time this misunderstanding created a barrier toward the implementation or adoption of a software engineering practice.
- Develop support using software tools during the standard training as well as for the implementation of the ISO/IEC 29110 standard. The experience of reinforce the 4 VSEs with the ISO/IEC 29110 show us that the implementation of this initiatives should have software tools that reduce the effort in the implementation of software engineering practices, so that, it helps to reduce the change resistance in the adoption and use of a new software engineering practices. However, this software tools should be developed in a way that they can be configured according to the VSE needs.

According to the experience of reinforcing the 4 VSEs with the ISO/IEC 29110, we can conclude that the set of practices provided in the standard can be easily imple-mented and help VSEs in providing quality products within approved budget and schedule.

Finally, as the next steps, we are working with more VSEs using agile method-ologies to get additional results about our implementing method to reinforce them with the practices provided by quality models or standards such as in the ISO/IEC 29110. We are also working to improve the efficiency and effectiveness of our five-step method.

# References

1. Muñoz, M., Mejia, J., Calvo-Manzano, J.A., San Feliu, T., Corona, B., Miramontes, J.: Diagnostic assessment tools for assessing the implementation and/or use of agile methodologies in SMEs: an analysis of covered aspects. Softw. Qual. Prof. **19**(2), 16–27 (20170). ISSN: 15220542
2. Sanchez-Gordon, M.-L., de Amescua, A., O'Connor, R.V., Larrueca, X.: A standard-based framework to integrate software work in small settings. Comput. Standars Interfaces **54**(Part 3), 117–194 (2017)

3. Larrucea, X., O'Connor, R. V., Colomo-Palacios, R., Laporte, C.Y.: Software process improvement in very small organizations. IEEE Softw. **33**(2), 85–89 (2016). https://doi.org/10.1109/ms.2016.42
4. Laporte, C., O'Connor, R.: Systems and software engineering standards for very small entities: accomplishments and overview. Comput. IEEE Comput. Soc. **49**(8), 84–87 (2016)
5. Laporte, C.Y., Muñoz, M., Gerançon, B.: The education of students about software engineering standards and their implementations in very small entities. In: IEEE Canada-International Humanitarian Technology Conference, July 20–21, Toronto, Ontario, Canada, pp. 94–98 (2017)
6. Chetankumar, P., Ramachandran, M.: Agile maturity model (AMM): a software process improvement framework for agile software development practices. Int. J. Softw. Eng. **2**(1), 3–28 (2009)
7. Beck, K., et al.: Manifesto for agile software development (2001). www.agilemanifesto.org/
8. Harleen, K., Flora, S., Chande, V.: A systematic study on agile software development methodologies and practices. Int. J. Comput. Sci. Inf. Technol. (IJCSIT) **5**(3), 3626–3637 (2014)
9. ISO. ISO/IEC TR 29110-5-1-1:2012- Software engineering – Lifecycle profiles for very small entities (VSEs) – Part 5-1-1: Management and engineering guide – Generic profile group: Entry profile. International Organization for Standardization/International Electrotechnical Commission, Geneva, Switzerland (2012). http://standards.iso.org/ittf/PubliclyAvailableStandards/c060389_ISO_IEC_TR_29110-5-1-1_2012(E).zip
10. Laporte, C., O'Connor, R., García Paucar, L.H.: The implementation of ISO/IEC 29110 software engineering standard and guides in very small entities. In: Maciaszek, L.A, Filipe, J. (eds.) ENASE 2015, CCIS 599, pp. 162–179. Springer, Cham (2016)
11. O'Connor, R.V., Laporte C.Y.: Software project management in very small entities with ISO/IEC 29110. In: Winkler, D., O'Connor, R.V., Messnarz, R. (eds.) EuroSPI 2012, CCIS 301, pp. 330–341. Springer, Berlin (2012)
12. Díaz, A., De Jesús, C., Melendez, K., Dávila, A.: ISO/IEC 29110 implementation on two very small software development companies in Lima. Lessons Learn. IEEE Lat. Am. Trans. **14**(5), 2504–2510 (2016)
13. Laporte, C., Muñoz, M., Mejia, J., O'Connor, R.: Applying software engineering standards in very small entities - from startups to grownups. IEEE Softw. **35**(1), 99–103 (2018)
14. NYCE, Companies certified in ISO/IEC 29110 standard 29110-4-1:2011 (2018). https://www.nyce.org.mx/wp-content/uploads/2018/02/PADRON-DE-EMPRESAS-CERTIFICADAS-ISO-IEC-29110-4-1.pdf

# Knowledge Management

# An Architecture for the Generation of Educational Rules – Based Games with Gamification Techniques

Humberto Marín-Vega[1](✉), Giner Alor-Hernández[1],
Luis Omar Colombo-Mendoza[1], Cuauhtémoc Sánchez-Ramírez[1],
Jorge Luis García-Alcaraz[2], and Liliana Avelar-Sosa[2]

[1] Tecnológico Nacional de México/I. T. Orizaba,
Av. Oriente 9, 852. Col Emiliano Zapata, 94320 Orizaba, Mexico
humbert_marin@outlook.com, galor@itorizaba.edu.mx,
colombo_isc@hotmail.com,
cuauhtemoc.sanchezr@gmail.com
[2] Universidad Autónoma de Ciudad Juárez,
Av. Plutarco Elias Calles 1210. Col. Foviste Chamizal,
31310 Ciudad Juárez, Chihuahua, Mexico
{jorge.garcia,liliana.avelar}@uacj.mx

**Abstract.** Gamification is the use of mechanics and designed techniques of games to involve and motivate people to achieve their goals. This work presents an architecture for the generation of educational rules-based games with gamification techniques. This architecture allows to develop gamification applications achieving learning objectives established by the user and the implementation of emerging technologies, these learning objectives are reflected through the use of learning activities and game attributes. As result of this work, a platform to generate educational rules-based games is presented, in this platform the layered architecture proposed is implemented with the use of gamification techniques.

**Keywords:** Architecture · Educational applications · Gamification
Rule based educational games · Serious games

## 1 Introduction

Gamification is "the use of design elements characterized for games in non-game contexts" [1]. It aims "to influence behavior, improve motivation and enhance compromise" [2]. Implementing gamification means adoption of rules-based games which are often complex, so it is much more practical to use software at least to track learners' achievements, make necessary calculations, and generate relevant feedback, rather than having the instructor manage all mentioned tasks. Gamification can take several forms, from the layering of basic game mechanics onto routine performance tracking, to the full integration of productive tasks into a virtual gaming environment [3]. A serious game is a "computer application, for which the original intention is to combine with consistency,

© Springer Nature Switzerland AG 2019
J. Mejia et al. (Eds.): CIMPS 2018, AISC 865, pp. 101–110, 2019.
https://doi.org/10.1007/978-3-030-01171-0_9

both serious aspects such as non-exhaustive and non-exclusive, teaching, learning, communication, or the information, with playful springs from a video game" [4].

Video games are a useful tool in teaching, video games allow active learning within a semiotic domain and provide critical knowledge, with learning and reflexing on the external grammar design of the games, which is a form of metacognition [5]. The aim of this work is to present a layered architecture and its components to generate gamified applications. The general idea behind the architecture is that it has to satisfy users with low and high requirements in the gamification aspect. As the first step, basic gamification mechanisms are available when configuring the game attributes. This architecture has a layered design to organize its components. Educational applications with gamification are generated by following this process with established characteristics through learning and game attributes. The gamification techniques implemented are motivation, action and reward, these techniques are defined in the attributes of the game. In the gamification the motivation is able to provoke a series of emotional relations that direct the user towards the accomplishment of a task or determined action by means of the mechanics of the game and the action technique consists in every event that the user have to develop to reach the objective. On the other hand, the reward technique consists in constantly reward actions. For example, a player may collect points every few seconds during gameplay.

The paper is organized as follows: in Sect. 2, a set of related works focused on the use of gamification techniques and serious games on educational applications is presented. In Sect. 3, the architecture for educational rules-based games development with gamification techniques is presented, this section includes the architecture and the design considerations and in Sect. 4 a case study about the generating educational rules-based games for a Web environment. Zeus is a platform to generate educational rules-based games implementing the architecture proposed. Finally, the last two sections include, the conclusion about this paper, followed by a brief presentation of future work.

## 2  Related Works

In recent years, several studies have been proposed with the aim of improving the development of educational games. Most of these research papers have been focused on the use of gamification in a variety of contexts. In this section, we present a set of related works focused on the use of gamification techniques and serious games on educational applications. These works have been grouped according to the kind of application that was developed: (1) Use of Serious Games on educational applications and (2) Use of Gamification techniques on educational applications.

### 2.1  Use of Gamification Techniques on Educational Applications

Robson et al. [6] studied how gamification improves the way companies attract customers and employees was analyzed by following a Business Horizons approach based on the gamification principles to attract new customers, encouraging the participation of the employees in gamification activities controlling each case even the unsuccessful experiences are included. Lubin [7] proposed that gamification offers a different didactic approach of designing an opportunity to attract people and make a situation or instruction relevant through the implementation of gamification in educational training environments. Simões et al. [8] found distinctive characteristics of good games, in order to understand what makes sense to apply in the teaching process, for this an online platform was used, which is a collaborative and social learning environment. The objective was to extract the best elements of the game to adapt and use them in the teaching process. Armastrong [9] defined the concept of gamification as a way to take advantage of motivation mechanisms to increase individual investment in a system, process or resource. One way to apply a motivation mechanism is through extrinsic motivation where you participate only because of external factors not related to the nature of the activity itself, such as the chance to win a prize or through the threat of punishment. Dicheva et al. [10] presented a study of published empirical research on the application of gamification to education which employs a systematic mapping design. Consequently, the author proposed a categorical structure to classify the results of the research based on the extracted topics discussed in the reviewed articles.

### 2.2    Use of Serious Games on Educational Applications

Raybourn [11] presented a new more effective and ascendable paradigm for training and education called Transmedia learning. Transmedia learning is defined as the scalable message system that represents a story or a basic experience that is developed from the use of multiple media, this learning paradigm offers a more effective use of serious games for training and education. Latorre et al. [12] presented the SELEAG® video game engine (learning in serious games) which is a multiplayer graphic adventure system, inspired by graphic adventures. This engine is used to implement the serious game TimeMesh®, designed to impart knowledge and skills in the areas of History and Geography. Boughzala et al. [13] defined that a serious game combines a serious intention with game rules and game objectives. Serious games are often considered technological applications that used games to engage people in an experience through which a learning objective or professional training is explored. Martins et al. [14] proposed that a way to help both professionals and patients was developing serious games aimed at motor rehabilitation in physical therapy sessions. In this sense, the authors proposed a modular Back Office system for the centralized management of one or more games intended for physical therapy. Peddycord-Liu et al. [15] presented a method that suggests curricular sequencing based on the prediction relationship between mathematical objectives. For this, the author applied an analysis of serious games for the curricular sequencing integrated to the curriculum with the mathematical game, ST Math.

# 3   An Architecture for Educational Rules – Based Games Development with Gamification Techniques

The educational rules – based games provide the context in terms of the challenges, goals and actions and how these are formalized in relation to game design. In that sense, rules may be characterized as constraints that limit the actions of the player [16]. Playing a serious game is an activity of improving content knowledge, skills and competencies in order to achieve learning outcomes. Games are structured in two ways: comprising rules and challenges for learner through emergence and progression [17]. The game designer has control over the sequence of the events or rules. Although, there are game rules that can be influenced or changed by player's actions [18]. Below the game attributes of an educational rules – based game are presented: (1) Scoring: the record of points or strokes made by the competitors in a game or match. (2) Moving: this attribute is involved in changing the location of the player in levels, worlds, etc. (3) Time Levels: is the player time manager. (4) Progress bars: this attribute can be used to show the progress or lost of live points and (5) Game instructions (including victory conditions): instructions that help the player to reach the conditions to accomplish the victory.

## 3.1   Architecture Design Considerations

The general idea behind the architecture is that it has to satisfy users with low and high requirements in the gamification aspect. The first step, basic gamification mechanisms are readily available when configuring game attributes, such as progress bars, scorings, timer levels, moving and game instructions. For this reason, all the elements of the classification of rules are presented in a game attribute repository. The second most important part is when the game attributes are selected and processed to continue with de application configuration and the rules-based game generation.

## 3.2   Architecture for the Generation of Educational Rules – Based Games

The architecture presented has a layered design to organize its components. This design allows scalability and easy maintenance, since the tasks and responsibilities of the tool are distributed evenly. Figure 1 presents the general architecture for the generation of educational rules – based games. Each component has a specific function explained as follows:

Figure 1. Represents the architecture for the generation of educational rules – based games, then describes the process for the generation of gamification applications:

(1) Login. Authentication component to access for the platform.
(2) Wizard to enter the functions in the platform. In this component, the XML-based configuration document that will be sent to the integration layer will be generated internally.
(3) Game attributes request this component receives the XML-based configuration file in the application layer and prepares it to send it to the presentation layer.

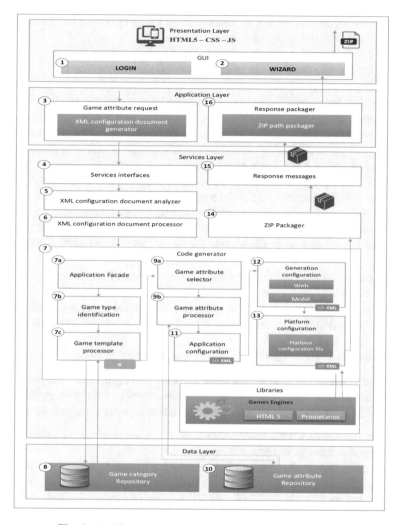

**Fig. 1.** Architecture of the rules games generation platform

(4) The service interfaces component receives the XML-based configuration document and sends it to the XML-based file analyzer.

(5) The XML-based configuration file is analyzed to check if it is well structured.

(6) The XML-based document is started to describe the application that will be generated.

(7) The code generator component is responsible for processing the XML-based configuration document and generates the application with gamification.

(7a) The application facade component uses the facade design pattern to provide a much simpler interface to communicate with the other components.

(7b) In this section, the game category that will be generated is identified.

(7c) The selected game category is processed and the template for this game category is consulted in the game category repository.

(8) The game categories repository stores the templates of each category, as well as the responses to requests for information.

(9a) In this part the game attributes that will contain the application are selected and generated, the selection consists first in selecting the attribute and then the type of that attribute.

(9b) The process of the attributes of the game consists in the generation of the game attributes previously selected. For its generation, the structure of the attributes is consulted in the game attribute repository.

(10) The repository of game attributes contains the description of the structure of the game attributes, as well as the description of its functionality.

(11) Description of the data of the application with the gamification.

(12) Generation selection. This section has two options to generate the game (1) Web to use in the browser or (2) Mobile for phones with Android™, iOS™ or Windows Phone™.

(13) The person on the platform receives the specifications file and prepares it to send it to the videogames engine (game engines) for processing.

(14) Recover the project generated in the game engine and generates a ZIP file.

(15) Take the ZIP file path and return a URL with the address that allows you to download the file.

(16) Returns the URL to the GUI to generate a button to download.

The architecture has a layered design in order to organize its components. The platform architecture for generation of educational applications with gamification is shown in Fig. 1 below, each of the layers that the architecture contains is explained:

– **Presentation layer:** This layer consists of providing interaction with the educational applications generation platform.
– **Application layer:** It is responsible for accessing the services layer, simplifying the service information by converting it to data that the interface understands.
– **Service layer:** It is responsible for providing a set of services (modules) offered by the tool. The service layer receives the XML-based configuration document and returns a compressed file with the generated application. The service layer communicates directly with the data layer to access the repositories.
– **Data layer:** The data layer stores the repositories where the information necessary for the development of the application with gamification is located.

### 3.3 XML-Based Configuration Document for Generating Educational Games

The XML-based configuration document is the file in which all the necessary information for the generation of the application with gamification is described, this file is generated on the platform as a tool that contains the application information, the game information (game engine, category and attributes) and the application configuration for the application with gamification, throughout the process generation all requirements of the user are described in the XML-based configuration document and this in turn is processed by different components of the platform to identify the user requirements and generate the application. In the Fig. 2 the graphic representation of the XML-based document is visualized, in which the different levels that make up the document are observed.

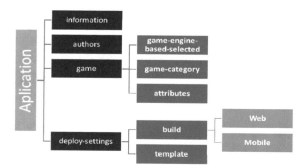

**Fig. 2.** Graphic representation of the XML-based schema

The XML-based configuration document contain elements, which are logical data structures that establish the configuration information. The configuration file is the result of the developing process of an educational application with gamification on the platform. To demonstrate the functioning of the architecture proposed, a platform for the generation of educational rule based games was developed, this platform is named Zeus, this platform is developed in PHP implementing XML and games engines for the generation of the game. Zeus is a serious game generator that facilitates the generation of serious games with a specific learning objective in base of learning activities and game attributes. The goal of Zeus is to assist the user to generate a serious game with the most appropriate attributes according to a game classifications and the learning activities that will use the game that should be generated.

## 4   Case Study: Generating Educational Rules – Based Games for a Web Environment on the Zeus Platform

In the present case study, the game rules used in support to basic education in a mathematics area was selected, because it is an area which presents difficulty for students to understand diverse topics. This due to diverse causes like the lack of interest, the ignorance of previous topics, among others. Mathematics is too wide-ranging of an area and that is why this studied case was defined on arithmetic. Arithmetic is the mathematics branch which study objectives are numbers and the elementary operations made with them, it is a study subject in basic education levels. On the Web platform for the development of applications with gamification, applications are generated to support the teaching - learning process in the area of elementary operations. The result for this case of study is a game to aid the learning of basic mathematics, the game generated is a set of rules which consists of the user catching a pointer which previously crashed with an object and depending on the object color it performs a sum of different scores at the player's scoring level. The learning attributes present in this application are the scoring, moving, timer levels, progress bar and game instructions including wining conditions. The process to generate a rules-based game is: the first step is to choose the selection of the type of game engine that will be implemented in the generation of the application. The options to be selected are HTML-based frameworks or proprietary frameworks (also called proprietary game engines), the next step and one of the most important, is the development of the game (Game configuration). This phase consists in selecting the attribute type that the rules-based game contains, this attribute is presented in a previous visualization according to its functionality like it can be observed in Fig. 3.

**Fig. 3.** Game attribute and game attribute types selection

The next step is the configuration part; in this section the user writes the important configuration data application to follow with the platform selection to generate the game. In this case, the generation of the application with gamification for the Web is presented, as shown in Fig. 4.

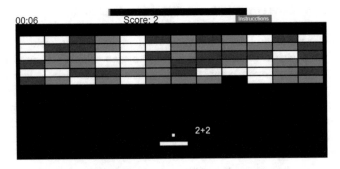

**Fig. 4.** Educational application with gamification

This game generated by the Zeus platform uses the gamification techniques of motivation, action and rewards. The action, this technique is comprised by the use of the moving game attribute which aims at the player's interaction with the objects that are presented in the game. In this case the action technique is that the player has to catch the pointer so that it does not fall into the emptiness and can bounce on it to address the objects at the top of the game, at the moment the player receives the pointer he adds points to the scoring, the values of the sum are depending on the object color with which the pointer collides. The player can move from left to right as long as he receives the pointer, when the player does not catch the pointer, points are subtracted shown in the progress bar. The motivation is worked through the game attributes themselves such as scoring, progress bar and timer levels. The scoring consists in increasing the score depending on the objects that the player catches, thus summing points depends on their success. In the progress bar, points are deducted when the player does not catch the pointer as time goes by motivating the player to catch the largest number of objects in the time set for the game. The timer levels are used to measure in which time you obtain a certain score. The interaction of these attributes motivates the user to obtain the greatest number of points in the least amount of time. The game of rules generated has a basic reward system in which each object has a score depending on the color of the object with which the pointer collides with. The main characteristic of the resulting applications is that gamification techniques are being implemented by learning activities and rules-based game attributes in order to achieve a learning objective.

## 5    Conclusions and Future Directions

Gamification is the use of game techniques and game mechanics in applications that are not a game as such, gamification is an innovative technique to assist the teaching-learning process. This work emerged from the need to develop educational applications with gamification that meets the set learning objectives. The architecture presented in this work aims to generate gamification applications which meet the learning objectives set by the user when developing their application, these objectives will be reflected through the use of the learning attributes that the user will select for their application. The architecture presented in this paper is implemented on the Zeus platform for the generation of educational rules - based games, the platform implementing the architecture with learning activities, game attributes and the gamification techniques of motivation, action and rewards. As future guidelines of this works, we will integrate more game categories proposed by Lameras on the Zeus platform to generate more educational applications with more gamification techniques, we shall propose to generate educational applications for web and mobile platforms (iOS and Android). Also, a quality-based attributes evaluation is being considered to carry out on the Zeus platform.

**Acknowledgements.** This work was sustained and sponsored by Tecnológico Nacional de México (TecNM), the National Council of Science and Technology (CONACYT), and the Secretary of Public Education (SEP) through PRODEP.

# References

1. Deterding, S., Dixon, D., Khaled, R., Nacke, L.: Gamification: from game design elements to gamefulness: defining "gamification". In: Proceedings of MindTrek 2011. ACM (2011)
2. Marczewski, A.: A response to Gartner's new definition of gamification (2014). https://www.gamified.uk/2014/04/05/a-response-to-gartners-new-definition-of-gamification/
3. Reeves, B., Read, J.L.: Total Engagement: Using Games and Virtual Worlds to Change the Way People Work and Businesses Compete. Harvard Business School Press, Boston (2009)
4. Djaouti, D., Alvarez, J., Jessel, J.-P.: Classifying serious games: the G/P/S model. In: Handbook of Research on Improving Learning and Motivation Through Educational Games: Multidisciplinary Approaches (2011). https://doi.org/10.4018/978-1-60960-495-0.ch006
5. Jiménez-Hernández, E.M., Oktaba, H., Piattini, M., Arceo, F.D.B., Revillagigedo-Tulais, A. M., Flores-Zarco, S.V.: Methodology to construct educational video games in software engineering. In: 2016 4th International Conference in Software Engineering Research and Innovation (CONISOFT), Puebla, pp. 110–114 (2016) https://doi.org/10.1109/conisoft.2016.25
6. Robson, K., Plangger, K., Kietzmann, J.H., Mccarthy, I., Pitt, L.: Game on: engaging customers and employees through gamification. Bus. Horiz. **59**(1), 29–36 (2016) https://doi.org/10.1016/j.bushor.2015.08.002
7. Lubin, L.: The gamification of learning and instruction field book. New Horiz. Adult Educ. Hum. Resour. Dev. **28**(1), 58–60 (2016)
8. Simões, J., Díaz-Redondo, R., Fernandez-Villas, A.: A social gamification framework for a K-6 learning platform. Comput. Hum. Behav. **29**, 245–253 (2013)
9. Armstrong, D.: The new engagement game: the role of gamification in scholarly publishing. Learn. Publ. **26**(4), 253–256 (2013)
10. Dicheva, D., Dichev, C., Agre, G., Angelova, G.: Gamification in education: a systematic mapping study. Educ. Technol. Soc. **18**(3), 75–88 (2015)
11. Raybourn, E.M.: A new paradigm for serious games: transmedia learning for more effective training and education. J. Comput. Sci. **5**(3), 471–481 (2014)
12. Andrés, P.L., Francisco, S., Moreno, J.L., de Carvalho, C.V.: TimeMesh, 1–8 (2014). https://doi.org/10.1145/2662253.2662353
13. Boughzala, I., Michel, H., Freitas, S.D.: Introduction to the serious games, gamification and innovation minitrack. In: 48th Hawaii International Conference on System Sciences. IEEE Computer Society (2015)
14. Martins, T., Carvalho, V., Soares, F.: Integrated solution of a back office system for serious games targeted at physiotherapy. Int. J. Comput. Games Technol. **2016** (2016). Article ID 1702051, 11 pages. https://doi.org/10.1155/2016/1702051
15. Liu, Z., et al.: MOOC learner behaviors by country and culture; an exploratory analysis. In: International Conference on Educational Data Mining (EDM), Raleigh, NC (2016)
16. Charsky, D.: From edutainment to serious games: a change in the use of game characteristics. Games Cult. **5** (2010). https://doi.org/10.1177/1555412009354727
17. Juul, J.: Half-Real. Video Games Between Real Rules and Fictional Worlds. The MIT Press, Cambridge (2005)
18. Lameras, P., Arnab, S., Dunwell, I., Stewart, C., Clarke, S., Petridis, P.: Essential features of serious games design in higher education: linking learning attributes to game mechanics. Br. J. Educ. Technol. **48**(4), 972–994 (2017)

# A Review on Bayesian Networks for Sentiment Analysis

Luis Gutiérrez, Juan Bekios-Calfa, and Brian Keith[✉]

Department of Computing and Systems Engineering,
Universidad Católica del Norte, Av. Angamos 0610, Antofagasta, Chile
{luis.gutierrez, juan.bekios, brian.keith}@ucn.cl

**Abstract.** This article presents a review of the literature on the application of Bayesian networks in the field of sentiment analysis. This is done in the context of a research project on text representation and use of Bayesian networks for the determination of emotions in the text. We have analyzed relevant articles that correspond mainly to two types, some in which Bayesian networks are used directly as classification methods and others in which they are used as a support tool for classification, by extracting features and relationships between variables. Finally, this review presents the bases for later works that seek to develop techniques for representing texts that use Bayesian networks or that, through an assembly scheme, allow for superior classification performance.

**Keywords:** Bayesian networks · Sentiment analysis · Literature review
Opinion mining

## 1 Introduction

Sentiment analysis, also known as opinion mining, is a type of natural language processing whose purpose is to perform the task of detecting, extracting, and classifying opinions, sentiments, and attitudes concerning different topics expressed in a text. Sentiment analysis helps in observing the attitudes of a population towards political movements [1], market intelligence [2], the level of consumer satisfaction with a product or service [3], box office prediction for feature films [4], among others [5, 6].

The availability of opinions and evaluations, in general, has increased in several fields, such as e-commerce [7], tourism [8] and the analysis of social networks such as Twitter [9], this has occurred together with the rise of Big Data. As an example, e-commerce consumers currently read product reviews published by previous consumers before purchasing, while producers and service providers improve their products and services by obtaining feedback from consumers.

Furthermore, there are numerous challenges that sentiment analysis must face [10]; for example, the same word may have negative connotations in some contexts, and positive in others; on the other hand, there is a great variety of ways in which people express their opinions, this means that small changes in syntax of the messages communicated can cause an important difference in the underlying opinions; this can be seen in the phrases "the movie was good" and "the movie was not good". Furthermore, the opinions expressed are not purely composed of a particular type of judgment, since

© Springer Nature Switzerland AG 2019
J. Mejia et al. (Eds.): CIMPS 2018, AISC 865, pp. 111–120, 2019.
https://doi.org/10.1007/978-3-030-01171-0_10

they may be composed of sentences that show a positive opinion on the subject, and others a negative one. All the above conditions, taken together, allow seeing the inherent difficulty in the task of sentiment analysis since this is challenging even for human beings [11].

The problems of sentiment analysis can be approached through various techniques, among them Bayesian networks. Bayesian networks are a modeling technique that allows describing dependency relationships between different variables by using a directed graph structure that encodes conditional probability distributions [12].

The aim of this work is to carry out a literature review regarding the application of Bayesian networks in the field of sentiment analysis. This review emerges within the context of an ongoing research project interested in providing, among other things, the development of adequate representation techniques for the modeling of emotions in text. In particular, this project contemplates the application of Bayesian probabilistic graphical models to perform or support classification tasks, by storing expert knowledge in the structure of these models. Thus, in order to prepare adequate theoretical foundations to do this, we need to explore the existing connection between sentiment analysis, natural language processing and Bayesian networks in the context of emotion determination in texts.

It should be noted that even though Naïve Bayes corresponds to the simplest Bayesian network model (a simple Bayesian network with a single root node) [13]. A large number of works in the field apply this technique to perform the classification task [6], thus, this technique has been extensively used and well-studied within the sentiment analysis literature. Considering this, for the purposes of this work Naïve Bayes has not been considered in the search, because the intention of this review is to analyze the more complex Bayesian Network structures and approaches that have been applied in sentiment analysis, in order to obtain better models than the simple Naïve Bayes approach. In this context, we have reviewed the sentiment analysis literature in order to find works that apply Bayesian approaches, from these works we have removed those that do not refer to any kind of Bayesian Network, which are the focus of our work, and we have removed those that are purely based on Naïve Bayes, since we decided to exclude it for the aforementioned reasons.

The rest of this document is structured as follows, in Sect. 2 a brief introduction to the basic concepts of Bayesian networks is given with the purpose of providing the necessary foundations to understand the rest of the work. Section 3 presents a review of the literature made with the different works found and their implications. Then, in Sect. 4 a discussion is made summarizing the main results that can be extracted from the review conducted. Finally, in Sect. 5 the conclusions of this work are provided.

## 2  Bayesian Networks

In this section, a brief introduction to Bayesian networks is given, using as a basis the exposition made by Crina and Ajith [12] among other references.

In the context of modeling and machine learning problems, Bayesian networks are typically used to find relationships among a large number of words. In this context, Bayesian networks provide an adequate tool to represent these relationships. These models are a type of probabilistic graphical model and in some cases, they are known as Bayesian belief networks. A Bayesian network consists of a directed acyclic graph where each node represents a random variable and the edges between the nodes represent an influence relationship (i.e., as the parent node influences the child node). These influences are modeled using conditional probability distributions [14–17].

A Bayesian network for a set of variables consists of a network structure that encodes conditional independence assertions about the variables, and a set of local probability distributions associated with each variable. Together, these two components allow defining a joint probability distribution for the set of all the problem variables. This conditional independence allows building a compact representation [12].

In order to define the conditional probability distributions, a table is defined (Conditional Probability Table or CPT). This table assigns probabilities to the variable of a node depending on the values of its parents in the graph. If a node does not have parents, the table assigns a probability distribution to that random variable [12]. An example of a Bayesian network is shown in Fig. 1. In this example, the variable of interest corresponds to an indicator of whether the text expresses a positive or negative sentiment, while the factors that determine this are given by the content of the text (variables indicating if the text contains "dog" and "love"). Note that if it is known a priori that event A is true, then the probability of event B changes, for this particular example the probability of occurrence of B increases from 0.3 to 0.6. Finally, in this example, the variable S depends both on events A and B as it can be seen in its CPT.

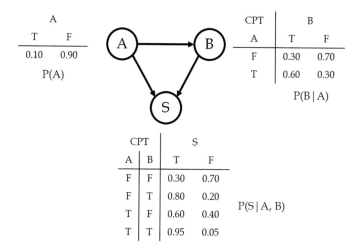

**Fig. 1.** Example of a simple Bayesian network (A: The text contains the word "dog"; B: The text contains the word "love"; S: The text expresses a positive sentiment).

The construction of a classifier using Bayesian networks requires first learning the structure of the network and their respective CPTs [18]. The concept of conditional probability tables can be extended to the continuous case in which variables can be based on other laws of probability such as a Gaussian distribution or solved by discretization [19–21].

One of the applications of Bayesian networks is that it is possible to make inferences about them using available information. In this context, there are several types of inference [22–26], such as diagnostic inference, causal inference, inter-causal inference, and mixed inferences. While inference in any Bayesian Network is an NP-hard problem [27], there are efficient alternatives that exploit conditional independence restrictions for certain kinds of networks [12, 28]. On the other hand, one of the benefits of Bayesian networks is that they can directly handle incomplete data sets (i.e., if one of their entries is missing) [12].

Bayesian networks have been applied successfully to solve different machine learning problems, even outperforming other popular machine learning approaches [29]. It should be noted that the simple Bayes classifier (Naïve Bayes) is a special case of a Bayesian network [30] and is one of the most used algorithms in the literature, especially in the context of sentiment analysis [6].

## 3    Literature Review

In the review carried out, two main tasks have been found in which Bayesian networks are used in sentiment analysis, specifically, these can be used to perform classification tasks directly or for the extraction of characteristics to support the classification task with another method. A brief description of the reviewed works is presented below, followed by a series of relevant observations extracted from them in the context of the project. The articles found are summarized in Table 1.

**Table 1.** Summary table of the reviewed articles by year.

| Year | Articles |
|------|----------|
| 2004 | [31] |
| 2011 | [32] |
| 2012 | [33, 34] |
| 2013 | [35, 36] |
| 2015 | [37] |
| 2018 | [42] |

Classification is one of the most recurrent uses of Bayesian networks. In this context, in Wan's work [37] and the article by Al-Smadi et al. [42] Bayesian networks are used directly as a sentiment classifier, obtaining competitive results and in some cases higher, when compared with other approaches.

In the work proposed by Chen et al. [43], the authors develop a parallel algorithm for Bayesian networks structure learning from large-scale text datasets. This algorithm is implemented by using a MapReduce cluster and is applied to capture dependencies among words. In particular, this approach allows us to obtain a vocabulary for extracting sentiments. The analysis of the experimental results obtained on a blogs dataset shows that the method is able to extract features with fewer predictor variables compared to the complete data set (i.e., a more parsimonious model), resulting finally in better predictions than the usual methods found in the literature.

The article by Lane et al. [34] addresses the problems of choice of models, feature extraction and unbalanced data in the field of sentiment analysis. The particular task that the authors address is classification, although they consider two different approaches, the first corresponds to classification of subjectivity (i.e., determining whether the text indicates something objective or subjective) and the second corresponds to determination of polarity (i.e., determining if the opinion of the text is positive or negative). In this paper, several classifiers and approaches to extracting features are evaluated, as well as the effect of balancing the data set before training. In some cases of the studied data sets, Bayesian networks showed a decrease in their performance when applying data balancing techniques, in contrast to the behavior of the other classifiers.

In the article by Ortigosa et al. [33] the authors address a multidimensional problem, in which they use three related dimensions for sentiment analysis. Most traditional approaches are focused on a one-dimensional case and are inappropriate, whereas multilabel approaches cannot be applied directly. Given this, the authors propose the use of a network of multidimensional Bayesian classifiers [44, 45]. In addition, applying semi-supervised techniques avoid the hard work of manually labeling the examples.

The work carried out by Airoldi et al. [31] and Bai [32] proposes a two-stage Markov Blanket Classifier to perform the task of extracting sentiments from unstructured text, such as film reviews, using Bayesian networks. This approach also uses the Tabu Search algorithm [46] to prune the resulting network and obtain better classification results. While this has proved useful in preventing the overfitting of models, their work does not fully exploit the dependencies among sentiments. In this context, the work carried out by Olubulu et al. [38] proposes an improvement in considering dependencies among different sentiments.

Another work by Olubolu [35] proposes an improvement for the Bayesian network model that includes sentiment-dependent penalties for the scoring functions of Bayesian networks (e.g., K2, Entropy, MDL and BDeu). The authors call this approach Sentiment Augmented Bayesian Network, which was evaluated and contrasted with the techniques used in the literature, obtaining competitive results and in some cases higher than the baseline. The proposed modification also derives the dependency structure of sentiments using conditional mutual information between each pair of variables in the data set. In a later work [29] this proposal is continued evaluating it in another domain (product reviews), here the knowledge contained in SentiWordNet [47]. The empirical results in the evaluated data sets suggest that this sentiment-dependent model could improve the classification results in some specific domains.

In the paper by Ren and Kang [36] a hierarchical approach is proposed for the modeling of simple and complex emotions in text. While most of the work focuses on the modeling of simple emotions in each document, the reality is that many documents are associated with complex human emotions that are a mixture of simple emotions difficult to model. The hierarchical model is superior to the traditional machine learning techniques used in the baseline (such as Naïve Bayes and Support Vector Machines) for the task of classifying simple emotions, while for complex emotions it shows promising results. The analysis of results also indicates that there is a relationship between the topics of the documents and the emotions contained in them.

The study of complex emotions is also addressed in the article by Wang et al. [39]. The authors evaluate multilabel sentiment analysis techniques on a set of data obtained from Chinese weblogs (Ren-CECps). Using the theory of probabilistic graphical models and Bayesian networks, the latent variables that represent emotions and topics are used to determine complex emotions from the sentences of weblogs. The analysis of the experimental results demonstrates the effectiveness of the model in recognizing the polarity of emotions in this domain.

In the work proposed by Chaturvedi et al. [40], a Bayesian network is used in conjunction with convolutional networks for the detection of subjectivity, this work is continued in [41]. The authors introduce a Bayesian Deep Convolutional Neural Network that has the ability to model higher-order features through several sentences in a document. One of the differences with other works is that they use Gaussian Bayesian networks to learn the features that are fed to the convolutional neural network. Their proposal delivers superior results to those of the different approaches used in the baseline.

## 4   Discussion

The tasks in which Bayesian networks are used are also recurrent in articles where machine learning techniques, such as Naïve Bayes and Support Vector Machines, are used. For this reason, it is expected that the performance of models based on Bayesian networks are also subject to, at least, some drawbacks that normally arise in environments where machine learning techniques are used. For example, the problem of high dimensionality in the domain is identified (e.g., sentiment analysis) and solutions are proposed, such as the extraction of a reduced vocabulary to perform classification tasks.

According to the reviewed literature, overfitting of the model poses another problem, which the reviewed articles solve by pruning the Bayesian network. On the other hand, the segmentation of the data set and its correct balancing also appears as a challenge, whose solution reports positive or negative variations depending on the specific domain in which Bayesian networks are applied.

Furthermore, another recurrent and relevant topic is sentiment analysis carried out on complex emotions; among these articles, the work of Chaturvedi et al. [40, 41] which approaches the problem through ensemble learning relying on the main technique of Bayesian networks and convolutional neural networks (CNN).

The applications proposed in the articles together with their respective scopes are varied, ranging from the use of simple Bayesian networks as direct classifiers to the incorporation of this model with other more complex deep learning ones [48]. Such versatility provides a perspective that is taken into account in the research project, since the range of application possibilities of Bayesian networks in sentiment analysis not only is broad but also is presented as a still extensible field.

One of the challenges of sentiment analysis is correctly addressing how context affects the semantic orientation of the words [11]. In this context, it should be noted that the approaches previously discussed do not directly address this challenge, but rather focus on either classification or feature extraction. Those centered on classification could benefit from using an adequate representation that addresses this issue, while those that are centered on feature extraction could possibly be modified to partially address this issue, through the addition of contextual information into the extracted features.

In the context of the project addressed, one of the work dimensions consists of the development of an encoding of the texts to obtain an adequate representation. In particular, the construction of a highly discriminating features space for the text is sought with the purpose of determining the emotions present in it (i.e., a sentiment analysis task). Following this line of work, the development of classifiers that have more than one output (multidimensional [49] or multilabel [50, 51] as appropriate) has been considered to obtain the different emotions of the text. To achieve this, the line of work of multidimensional Bayesian classifiers is considered, these methods based on graphical models allow to support the multidimensional classification by storing expert knowledge in its structure. The literature review indicates that Bayesian networks have been used with varying levels of success, but that there are still several challenges regarding their application [52–54].

In particular, with regards to the objective of finding an adequate representation of text for the task of determining emotion, the works by Chen et al. [43], Ortigosa et al. [33] and Chaturvedi et al. [40] are the most interesting in the context of this project. The first work due to its focus on feature extraction using fewer prediction variables, the second article because of its focus on multidimensional sentiment analysis, which is well-aligned with the objectives of the project, and finally the third paper because of its use of deep learning in order to model higher-order features, which could be useful in our search of an adequate representation.

# 5   Conclusions

In this work, a literature review on the application of Bayesian networks in the field of sentiment analysis has been conducted. In this context, the work's objectives are considered accomplished. It should be noted again that the applications of Bayesian networks are varied, although the main ones are as a classification method or as a support tool, by extracting features and relationships between variables. As part of future work, we seek to develop text representation techniques based on the concepts shown above. As mentioned above, sentiment analysis, and in particular the modeling of emotions, has a wide range of applications, so that an adequate representation could benefit them. In addition, subsequent studies can be conducted to find models, additional to CNN, that can be assembled with Bayesian networks in such a way that classification performance improves maintaining a favorable trade-off.

**Acknowledgments.** Research partially funded by the National Commission of Scientific and Technological Research (CONICYT) and the Ministry of Education of the Government of Chile. Project REDI170607: "Multidimensional Bayesian classifiers for the interpretation of text and video emotions".

# References

1. Tumasjan, A., Sprenger, T.O., Sandner, P.G., Welpe, I.M.: Predicting elections with twitter: what 140 characters reveal about political sentiment. In: ICWSM, vol. 10, no. 1, pp. 178–185 (2010)
2. Li, Y.M., Li, T.Y.: Deriving market intelligence from microblogs. Decis. Support Syst. **55** (1), 206–217 (2013)
3. Ren, F., Quan, C.: Linguistic-based emotion analysis and recognition for measuring consumer satisfaction: an application of affective computing. Inf. Technol. Manag. **13**(4), 321–332 (2012)
4. Nagamma, P., Pruthvi, H., Nisha, K., Shwetha, N.: An improved sentiment analysis of online movie reviews based on clustering for box-office prediction. In: 2015 International Conference on Computing, Communication and Automation (ICCCA), pp. 933–937. IEEE (2015)
5. Liu, B.: Web Data Mining: Exploring Hyperlinks, Contents, and Usage Data. Springer Science & Business Media (2011)
6. Ravi, K., Ravi, V.: A survey on opinion mining and sentiment analysis: tasks, approaches and applications. Knowl. Based Syst. **89**, 14–46 (2015)
7. Akter, S., Wamba, S.F.: Big data analytics in e-commerce: a systematic review and agenda for future research. Electron. Mark. **26**(2), 173–194 (2016)
8. Alaei, A.R., Becken, S., Stantic, B.: Sentiment analysis in tourism: capitalizing on big data. J. Travel. Res. (2017). https://doi.org/10.1177/0047287517747753
9. Sehgal, D., Agarwal, A.K.: Real-time sentiment analysis of big data applications using Twitter data with Hadoop framework. In: Soft Computing: Theories and Applications, pp. 765–772. Springer (2018)
10. Cambria, E., Das, D., Bandyopadhyay, S., Feraco, A.: A practical guide to sentiment analysis, vol. 5. Springer (2017)
11. Liu, B.: Sentiment analysis and opinion mining. Synthesis Lectures on Human Language Technologies, vol. 5, no. 1, pp. 1–167 (2012)
12. Grosan, C., Abraham, A.: Intelligent systems. Springer (2011)
13. Mononen, T., Myllymäki, P.: Fast NML computation for Naive Bayes models. In: International Conference on Discovery Science, pp. 151–160. Springer (2007)
14. Kass, R.E., Raftery, A.E.: Bayes factors. J. Am. Stat. Assoc. **90**(430), 773–795 (1995)
15. Jensen, F.V.: An Introduction to Bayesian Networks, vol. 210. UCL Press, London (1996)
16. Heckerman, D., Geiger, D., Chickering, D.M.: Learning Bayesian networks: the combination of knowledge and statistical data. Mach. Learn. **20**(3), 197–243 (1995)
17. Bernardo, J.M., Smith, A.F.: Bayesian Theory (2001)
18. Cooper, G.F., Herskovits, E.: A Bayesian method for the induction of probabilistic networks from data. Mach. Learn. **9**(4), 309–347 (1992)
19. John, G.H., Langley, P.: Estimating continuous distributions in Bayesian classifiers. In: Proceedings of the Eleventh Conference on Uncertainty in Artificial Intelligence, pp. 338–345. Morgan Kaufmann Publishers Inc. (1995)

20. Driver, E., Morrell, D.: Implementation of continuous Bayesian networks using sums of weighted Gaussians. In: Proceedings of the Eleventh Conference on Uncertainty in Artificial Intelligence, pp. 134–140. Morgan Kaufmann Publishers Inc. (1995)
21. Friedman, N., Goldszmidt, M., et al.: Discretizing continuous attributes while learning Bayesian networks. In: ICML, pp. 157–165 (1996)
22. Ding, J.: Probabilistic inferences in Bayesian networks. arXiv preprint arXiv:1011.0935 (2010)
23. Wellman, M.P., Henrion, M.: Explaining 'explaining away'. IEEE Trans. Pattern Anal. Mach. Intell. **15**(3), 287–292 (1993)
24. Zhi-Qiang, L.: Causation, Bayesian networks, and cognitive maps. Acta Autom. Sin. **27**(4), 552–566 (2001)
25. Milho, I., Fred, A., Albano, J., Baptista, N., Sena, P.: An auxiliary system for medical diagnosis based on Bayesian belief networks. In: Proceedings of 11th Portuguese Conference on Pattern Recognition, RECPAD (2000)
26. Mori, J., Mahalec, V.: Inference in hybrid Bayesian networks with large discrete and continuous domains. Expert Syst. Appl. **49**, 1–19 (2016)
27. Cooper, G.F.: The computational complexity of probabilistic inference using Bayesian belief networks. Artif. Intell. **42**(2–3), 393–405 (1990)
28. Heckerman, D.: A tutorial on learning with Bayesian networks. In: Learning in Graphical Models, pp. 301–354. Springer (1998)
29. Orimaye, S.O., Pang, Z.Y., Setiawan, A.M.P.: Learning sentiment dependent Bayesian Network classifier for online product reviews. Informatica **40**(2), 225 (2016)
30. Cheng, J., Greiner, R.: Comparing Bayesian network classifiers. In: Proceedings of the Fifteenth Conference on Uncertainty in Artificial Intelligence, pp. 101–108. Morgan Kaufmann Publishers Inc. (1999)
31. Airoldi, E., Bai, X., Padman, R.: Markov blankets and meta-heuristics search: sentiment extraction from unstructured texts. In: International Workshop on Knowledge Discovery on the Web, pp. 167–187. Springer (2004)
32. Bai, X.: Predicting consumer sentiments from online text. Decis. Support Syst. **50**(4), 732–742 (2011)
33. Ortigosa-Hernández, J., Rodríguez, J.D., Alzate, L., Lucania, M., Inza, I., Lozano, J.A.: Approaching sentiment analysis by using semi-supervised learning of multi-dimensional classifiers. Neurocomputing **92**, 98–115 (2012)
34. Lane, P.C., Clarke, D., Hender, P.: On developing robust models for favourability analysis: model choice, feature sets and imbalanced data. Decis. Support Syst. **53**(4), 712–718 (2012)
35. Orimaye, S.O.: Sentiment augmented Bayesian network. In: Data Mining and Analytics 2013 (AusDM 2013), p. 89 (2013)
36. Ren, F., Kang, X.: Employing hierarchical Bayesian networks in simple and complex emotion topic analysis. Comput. Speech Lang. **27**(4), 943–968 (2013)
37. Wan, Y., Gao, Q.: An ensemble sentiment classification system of Twitter data for airline services analysis. In: 2015 IEEE International Conference on Data Mining Workshop (ICDMW), pp. 1318–1325. IEEE (2015)
38. Orimaye, S.O., Pang, Z.Y., Setiawan, A.M.P.: Towards a sentiment dependent Bayesian network classifier for online product reviews (2016)
39. Wang, L., Ren, F., Miao, D.: Multi-label emotion recognition of weblog sentence based on Bayesian networks. IEEJ Trans. Electr. Electron. Eng. **11**(2), 178–184 (2016)
40. Chaturvedi, I., Cambria, E., Poria, S., Bajpai, R.: Bayesian deep convolution belief networks for subjectivity detection. In: 2016 IEEE 16th International Conference on Data Mining Workshops (ICDMW), pp. 916–923. IEEE (2016)

41. Chaturvedi, I., Ong, Y.S., Tsang, I.W., Welsch, R.E., Cambria, E.: Learning word dependencies in text by means of a deep recurrent belief network. Knowl. Based Syst. **108**, 144–154 (2016)
42. Al-Smadi, M., Al-Ayyoub, M., Jararweh, Y., Qawasmeh, O.: Enhancing aspect-based sentiment analysis of Arabic hotels' reviews using morphological, syntactic and semantic features. Inf. Process. Manag. (2018). https://www.sciencedirect.com/science/article/abs/pii/S0306457316305623
43. Chen, W., Zong, L., Huang, W., Ou, G., Wang, Y., Yang, D.: An empirical study of massively parallel Bayesian networks learning for sentiment extraction from unstructured text. In: Asia-Pacific Web Conference, pp. 424–435. Springer (2011)
44. Van Der Gaag, L.C., De Waal, P.R.: Multi-dimensional Bayesian network classifiers (2006)
45. Bielza, C., Li, G., Larranaga, P.: Multi-dimensional classification with Bayesian networks. Int. J. Approx. Reason. **52**(6), 705–727 (2011)
46. Glover, F., Laguna, M.: Tabu search. In: Handbook of Combinatorial Optimization, pp. 3261–3362. Springer (2013)
47. Baccianella, S., Esuli, A., Sebastiani, F.: SentiWordNet 3.0: an enhanced lexical resource for sentiment analysis and opinion mining. In: LREC, vol. 10, pp. 2200–2204 (2010)
48. Wang, H., Yeung, D.Y.: Towards Bayesian deep learning: a survey. arXiv preprint arXiv: 1604.01662 (2016)
49. Zaragoza, J.H., Sucar, L.E., Morales, E.F., Bielza, C., Larranaga, P.: Bayesian chain classifiers for multidimensional classification. IJCAI **11**, 2192–2197 (2011)
50. Tsoumakas, G., Katakis, I.: Multi-label classification: an overview. Int. J. Data Warehous. Min. (IJDWM) **3**(3), 1–13 (2007)
51. Sucar, L.E., Bielza, C., Morales, E.F., Hernandez-Leal, P., Zaragoza, J.H., Larrañaga, P.: Multi-label classification with Bayesian network-based chain classifiers. Pattern Recogn. Lett. **41**, 14–22 (2014)
52. Drury, B., Valverde-Rebaza, J., Moura, M.F., de Andrade Lopes, A.: A survey of the applications of Bayesian networks in agriculture. Eng. Appl. Artif. Intell. **65**, 29–42 (2017)
53. Weber, P., Medina-Oliva, G., Simon, C., Iung, B.: Overview on Bayesian networks applications for dependability, risk analysis and maintenance areas. Eng. Appl. Artif. Intell. **25**(4), 671–682 (2012)
54. Wiegerinck, W., Burgers, W., Kappen, B.: Bayesian networks, introduction and practical applications. In: Handbook on Neural Information Processing, pp. 401–431. Springer (2013)

# Identifying the Orientations of Sustainable Supply Chain Research Using Data Mining Techniques: Contributions and New Developments

Carlos Montenegro[1(✉)], Marco Segura[1], and Edison Loza-Aguirre[1,2]

[1] Escuela Politécnica Nacional, Quito, Ecuador
{carlos.montenegro,marco.segura,
edison.loza}@epn.edu.ec
[2] CERAG FRE 3748 CNRS/UGA, 150, rue de la Chimie, BP 47,
38040 Grenoble Cedex 9, France
lozaedison@univ-grenoble-alpes.fr

**Abstract.** To be effective, sustainable development must maintain an equilibrium between its social, environmental and economic efforts. Several studies have suggested that an unbalance exists about the attention given to those three dimensions; however, few contributions have demonstrated such unbalance. This research describes a synthesis of two manual and semiautomatic methods published in the technical literature and includes additional developments, conceived to speed up and increase the accuracy of the analysis of the sustainable orientation of a corpus. The results are compared with the previous studies on about ten years of literature from top-tier journals dealing with Sustainable Supply Chain issues. The results confirm unbalance on research in this field. They show that most of the studies have been focussed on environmental and economic aspects, leaving aside social issues.

**Keywords:** Sustainable Supply Chain · LDA topic model · Feature selection
Content analysis · Sustainable Development

## 1 Introduction

In recent years, Sustainable Development (SD) has become an area of knowledge with multiple applications for the public and private sectors, calling the attention of researchers around the world. The first publications which combined SD with logistics and supply chain management appeared in the early 2000's originating the notion of Sustainable Supply Chain (SSC).

Since SSC can be considered an emergent research area, a general framework for its application has not been agreed upon yet [1, 2]. In this context, Green Supply Chain Management (GSCM) was born as a combination of environmental management and supply chain management [3], emphasizing the idea that an economic profit can be originated from activities oriented to reduce a possible ecological impact [4].

Later in this process of evolution, a new alternative known as Logistics Social Responsibility (LSR) was introduced by researchers, covering the need of considering

© Springer Nature Switzerland AG 2019
J. Mejia et al. (Eds.): CIMPS 2018, AISC 865, pp. 121–131, 2019.
https://doi.org/10.1007/978-3-030-01171-0_11

the social aspects of SD, which have been left aside by GSCM. LSR is based on the concepts of Corporate Social Responsibility giving high importance to both social and environmental issues, but less attention to economic aspects, reducing the ability of social or environmental initiatives to be sustainable over time without economic success [5, 6].

To remediate the limitations of LSR, introducing economic issues into the SD environment, SSC Management was proposed as "the management of material, information, and capital flow as well as cooperation among companies along the supply chain while taking goals from all three dimensions of SD into account which are derived from customer and stakeholder requirements" [7]. This definition of SSC is based on the principles of Elkington's triple bottom line [8], and clearly stated the need of a balance all three dimensions of SD (economic, environmental and social).

Even though the need of a balance between the three dimensions of SD has been recognized as the most desirable scenario, researchers have identified an unbalance on the consideration given to the three dimensions of SD [2, 4]. In this scenario and building on previous contributions [9, 10] which highlighted unbalance on SD research, the current research synthesizes the existent approaches and pose new developments in the analysis, using data mining techniques.

## 2  Base Contributions About Research Orientation on SSC

Previous research has proposed mechanisms designed to evaluate sustainability initiatives in organizations [11–13]. However, only a few researchers have analyzed the dimensional orientation of sustainable initiatives.

Loza-Aguirre et al. [9] proposed a tool to graphically represent the concentration of scientific publications along each one of the SD dimensions (social, economic or environmental, or any intersection of them). To achieve this goal, the authors collected about ten years of SSC literature from top-tier journals in the fields of Supply Chain Management and Operations. Then, these articles were manually coded to identify their orientation over the three dimensions of SD. Finally, and based on Elkington's triple bottom line for representing SD [8], the authors used a Monte Carlo approach to calculate and describe the dimensions of SD graphically as circular areas which size is proportional to the number of articles coded on each dimension. The results of this initiative [9] indicate that: (1) most of the research in SSC is oriented to the intersection of environmental and economic aspects, and (2) from the three dimensions of SD, social issues are the less studied.

In contrast, Montenegro et al. [10] develop an alternative approach which combines a probabilistic topic model and an expert intervention. Working on the same corpus, the results confirmed the unbalance on research in the field of SD. These results validated that most of the academic contributions are oriented to topics combining both environmental and economic aspects, giving less attention to social issues.

Figure 1 shows the visual representations of Elkington's triple bottom line [8], and the results reported by Loza-Aguirre et al. and Montenegro et al. which are presented on the right side with the same image since both have a similar geometric form.

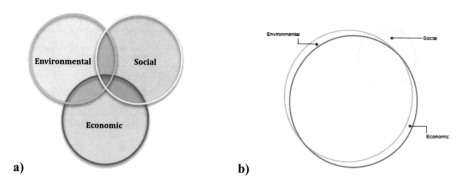

**Fig. 1.** (a) The triple bottom line for representing SD [8]; (b) Visual representation of alternative efforts [9, 10]

## 3 Data Collection

This research analyses the same set of articles as [9] and [10]. Table 1 presents the journals cited by the authors and the corresponding search queries.

**Table 1.** Journals consulted and searched queries used to collect corpus.

| Journals | Queries |
|---|---|
| Inter. J. of Logistics Management | Sustainable AND supply chain |
| Inter. J. of Logistics: Research and | Sustainable AND logistics |
| Applications | Green AND supply chain |
| Inter. J. of Operations and Production | Green AND logistics |
| Management | Sustainability AND supply chain |
| Inter. J. of Physical Distribution & Logistics | Sustainability AND logistics |
| Management | Social AND sustainable AND supply chain |
| Inter. J. of Production Economics | Social AND sustainable AND logistics |
| Inter. J. of Production Research | Social AND sustainability AND supply chain |
| J. of Business Logistics | Social AND sustainability AND logistics |
| J. of Operations Management | Social AND responsibility AND supply chain |
| J. of Supply Chain Management | Social AND responsibility AND logistics |
| Production and Operations Management | |
| Production Planning and Control | |
| Supply Chain Management | |
| Transportation Research Part E | |

The process carried out obtained 193 articles. Each article was carefully read and coded according to the principal orientation of their problematic. For coding, it was used the coding scheme presented in Table 2.

**Table 2.** SD orientation of documents through a Coding scheme

| SD dimension | | |
| --- | --- | --- |
| Environment | Social | Economic |
| Energy conservation | Ethics | Cost reductions |
| Environmental friendly transport | Human rights | Sustainable competitive advantage |
| Environmental measurement and evaluation | Philanthropy | Financial revenues |
| Gas emissions reduction | Quality of life | |
| Green manufacturing | Local communities | |
| Green product design | Safety | |
| Green purchasing and supply | Diversity | |
| Green urban logistics | | |
| Reverse logistics | | |
| Waste management | | |

## 4   Methods

With the purpose of coding the corpus of documents, this research employs the Latent Dirichlet Allocation (LDA) topic model, which generates the Topics from Bag of Words (BoW) codification, where the sequence of words has a probability which is not affected by the order in which they appear [14].

Then, an option to classify the documents (articles) is to use an unsupervised learning technique with an artificial neural network. Next, it is possible to filter the importance of the Topics, using a feature selection technique.

The classified documents allow to identify the relationships of SD dimensions, where the Topics of the documents are the content unit, and the final selected topics show the importance of them in the corpus. Figure 2 shows the process.

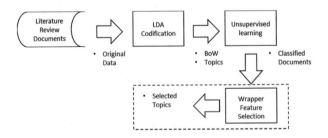

**Fig. 2.** The process to identify relevant topics of the corpus

## 4.1 Feature Selection and Classifiers

A feature is an individual measurable property of the process being observed, represented by a variable. Feature Selection (variable elimination) helps in understanding data, reducing computation requirement, reducing the effect of the dimensionality and improving the predictor performance [15–17].

The availability of label information allows the use of supervised feature selection algorithms [18, 19]. Based on the different strategies of searching, the Wrappers methods involve a learning algorithm as a black box and consist of using its prediction performance to assess the relative usefulness of subsets of variables. That is, the supervised feature selection algorithm uses the learning method (Classifier) as a subroutine with the computational burden that comes from calling the learning algorithm to evaluate each subset of features [20, 21], as presented in Fig. 3.

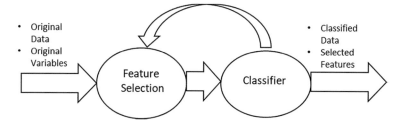

**Fig. 3.** Wrapper method [21]

To apply the described process, it is necessary to use two classifiers and a search method. Regarding the unsupervised classifier, this paper uses a Self-organizing neural network, with competitive neural network version, which can detect regularities and correlations in the input and adapt their future responses to that input accordingly [22].

Competitive learning models [22, 23] are based on the concept of the winning neuron, defined as the one whose weight vector is the closest to the current input vector, according to the distance (Euclidean in this case). During the learning phase, the weights of the winning neurons are modified incrementally to extract average features from the input patterns.

On the other hand, in a preliminary test round, using the classification accuracy as performance measure [24], the best results were shown by Bayes Net as wrapper method and Greedy Stepwise as search method.

The greedy stepwise performs a greedy forward or backward search through the space of attributes subsets [25]. On the Bayes Net algorithm, every attribute (every leaf in the network) is independent of the rest of the characteristics [26]. According to Cheng and Greiner [27], Bayesian Network classifiers are easy to construct, and the classification process is efficient.

## 5    Results

The results of the coding process conducted by Experts are presented in Table 3.

**Table 3.** Number of articles by category resulting from the coding.

| Totals by category | | % |
|---|---|---|
| Total of Economic articles | 21 | 0,1088 |
| Total of Social articles | 8 | 0,0414 |
| Total of Environmental articles | 7 | 0,0362 |
| Total of Economic and Social shared articles | 9 | 0,0466 |
| Total of Economic and Environmental shared articles | 123 | 0,6373 |
| Total of Social and Environmental shared articles | 5 | 0,0259 |
| Total of shared components among the three articles | 20 | 0,1036 |

To conduct the LDA implementation, it is used the Gibbs sampling [28], a form of Markov Chain Monte Carlo, which provides a relatively efficient method for extracting a set of topics from a corpus. Since there does not exist an optimal solution to estimate the number of topics (K), it is followed a best practice [29] which combine maximization [30, 31] and minimization [32, 33] for determining K as the number which best suits the considered models. By following this approach, the estimated K for the corpus was 42.

An excerpt of the numbered topics in order of appearance, their probabilities, and their top words are presented in Table 4. For the calculation, this research used 2000 iterations and recommended values of $\alpha$ (50/K) and $\beta$ (0.01) [34]. The solution was implemented in R, using related libraries. For preprocessing, comparison and testing it is used the MATLAB LDA implementation.

Based on their top words, each calculated topic of Table 4 was manually classified by an expert, according to the dimension of SD that better correspond to the top words of each topic (e.g., environmental, economic, social). Then, the sum the probabilities of the topics for each category is the aggregated probability or 'weight' of each dimension or intersection of dimensions. This value constitutes the SD dimensions using LDA model combined with the human expert participation. Note that in this case the Topics and their top words facilitate the expert work because diminish the number of items to analyze.

Alternatively, the documents, codified through the frequencies of the topic words, are classified using the Competitive model. For data processing, *nntool* of MATLAB [22] was used with 500 epochs and a learning rate of 0,1. The logical number of categories is seven, as the number of SD dimensions and intersections (see Table 5). The column "Competitive" is the normalized frequency of the number of the Articles for the automatized process.

The classified data can be used to select the relevant variables (features) using Greedy Stepwise with Bayes Net Algorithm, which is implemented in Weka [35, 36]. The combination has a significant value of Merit [36] of the best subset found.

**Table 4.** Calculated topics for the corpus and their probability (Excerpt)

| Topic 11 0.30178 | Topic 8 0.08159 | Topic 22 0.03764 | Topic 26 0.03420 | Topic 1 0.03061 | Topic 34 0.03017 | Topic 41 0.02861 |
|---|---|---|---|---|---|---|
| environmental | cost | environmental | product | sustainability | gscm | environmental |
| supply | costs | green | products | social | practices | plant |
| can | model | performance | remanufacturing | sustainable | environmental | suppliers |
| also | inventory | practices | reverse | companies | performance | management |
| management | time | suppliers | recovery | economic | pressures | supply |
| research | total | companies | life | firm | chinese | collaboration |
| chain | transportation | management | manufacturing | firms | china | Pollution |

| Topic 7 0.02842 | Topic 20 0.02791 | Topic 25 0.02786 | Topic 6 0.02504 | Topic 23 0.02280 | Topic 42 0.02269 | Topic 40 0.01936 |
|---|---|---|---|---|---|---|
| psr | transport | supply | supplier | logistics | csr | knowledge |
| purchasing | freight | chain | factors | reverse | suppliers | model |
| research | vehicle | chains | decision | university | social | sharing |
| managers | road | firms | suppliers | articles | companies | relational |
| lsr | logistics | green | performance | journal | practices | firm |
| carter | vehicles | strategic | selection | research | responsibility | green |
| activities | emissions | theory | evaluation | management | supplier | relationships |

| Topic 37 0.01877 | Topic 30 0.01674 | Topic 32 0.01646 | Topic 5 0.01239 | Topic 14 0.01171 | Topic 13 0.01150 | Topic 38 0.01129 |
|---|---|---|---|---|---|---|
| product | organizations | activity | green | recycling | time | given |
| csr | investments | disassembly | bargaining | packaging | emissions | nuclear |
| demand | suppliers | cost | producer | waste | vehicle | chain |
| activities | supplier | model | manufacturer | container | policy | logistics |
| levels | sustainable | treatment | greening | materials | window | waste |
| period | practices | can | supplier | information | per | reverse |

**Table 5.** Number of articles for the SD dimensions

| No. | SD Dim | Number of articles | Competitive |
|---|---|---|---|
| 1 | Economic + Social | 1 | 0,00518 |
| 2 | Environmental + Social | 3 | 0,01554 |
| 3 | Economic + Environmental + Social | 7 | 0,03627 |
| 4 | Only Social | 16 | 0,08290 |
| 5 | Only Environmental | 28 | 0,14507 |
| 6 | Only Economic | 30 | 0,15544 |
| 7 | Environmental + Economic | 108 | 0,55959 |
| | Total | 193 | |

Variables Topic 8, Topic 11, Topic 20, Topic 23, Topic 26, Topic 30, and Topic34, are selected as representatives of the Corpus. The group of selected Topics evidence that the feature selection does not coincide with the probabilities. Nevertheless, the chosen topics are in the group of the significant probabilities (see resalted Topics in Table 4).

The results of the three techniques, described and analyzed in this study, are displayed in Table 6, for each dimension of SD.

Finally, the graphical results are presented in Fig. 4. The results are obtained using the same Monte Carlo based technique, as in [9].

**Table 6.** Data of SD Dimensions to the graphical representation

| No. | SD Dim | Competitive | LDA + Expert | Expert |
|---|---|---|---|---|
| 1 | Economic + Social | 0,00518 | 0,07633 | 0,0466 |
| 2 | Environmental + Social | 0,01554 | 0,01828 | 0,0259 |
| 3 | Economic + Environmental + Social | 0,03627 | 0,06640 | 0,1036 |
| 4 | Only Social | 0,08290 | 0,03880 | 0,0414 |
| 5 | Only Environmental | 0,14507 | 0,07002 | 0,0362 |
| 6 | Only Economic | 0,15544 | 0,16820 | 0,1088 |
| 7 | Environmental + Economic | 0,55959 | 0,56198 | 0,6373 |

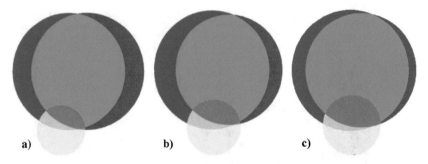

a)                    b)                    c)

**Fig. 4.** Relationships of SD dimensions according to the models: (a) Competitive; (b) LDA + expert; (c) Expert. Color significance: Blue for environmental; Red for economic; Orange for social. Other colors represent the correspondent combinations. Example: Green for social + environmental

It can be seen there are little graphical differences between the approaches, and the general tendency is the same; that is, the unbalance on research in SSC field, and it is corroborated that most of the academic contributions are oriented to topics combining both environmental and economic aspects, leaving aside social themes.

Additionally, using the data of Table 6, Fig. 5 shows the frequency (probability) of the SD dimensions, obtained manually (Expert), semi automatized (Expert plus LDA codification), and automatized (Competitive classification). The t-Test probe that the results have the same mean, at least with a 95% of probability.

The corpus preprocessing time is the same in all cases because the manual approach requires the use of statistical tools for the content analysis. The semimanual and automatized approaches use the preprocessed corpus. Nevertheless, the time difference is evident to identify the topics: weeks for the manual procedure, days for the semi-manual procedure and hours for the automatized process.

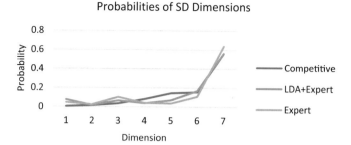

**Fig. 5.** Probabilities of the calculated SD dimensions

## 6  Discussion and Conclusion

Here it is proposed an alternative method which uses LDA probabilistic topic model and features selection instead of manual or semimanual analysis applied to classify sustainable efforts according to their SD orientation. The results obtained are compared with those from a manual and semimanual coding procedure performed on about ten years of literature from top-tier journals dealing with SSC issues [9, 10]. Working on the same corpus as [9], the results confirm unbalance on research in this field. It is validated that most of the academic contributions are oriented to topics combining both environmental and economic aspects, leaving aside social issues. However, the difference in time dedicated for each analysis is evident. Thus, apart from the results concerning sustainable orientation on SSC research, this paper is another example of the validity of data mining models as an alternative for analyzing a corpus.

The main originality of this contribution lies in the synthesis of the different models to analyze the Corpus of technical information about SSC theme, deriving in improvements and developments of an automatized process, including feature selection techniques and an unsupervised learning technique using neural networks. The scope of automatized corpus analysis can be extended; for example, besides the most representative topics, to identify the most influential articles and authors of the corpus.

The automatized approach would allow, on further research, the analysis of sustainability orientation on corpora in other fields. The method complements the available toolbox for analyzing the direction of the initiatives. It can be used not only in academia but also by governmental agencies or private organizations.

## References

1. Carter, C., Easton, P.: Sustainable supply chain management: evolution and future directions. Int. J. Phys. Distr. Log. **41**(1), 46–62 (2011)
2. Pagell., M., Shevchenko, S.: Why research in sustainable supply chain management should have no future. J. Supply Chain. Manag. **50**(1), 44–51 (2014)
3. Srivastava, S.: Green supply-chain management: a state-of-the-art literature review. Int. J. Manag. Rev. **9**(1), 53–80 (2007)

4. Carter, C., Rogers, D.: A framework of sustainable supply chain management: moving toward new theory. Int. J. Phys. Distr. Log. **38**(5), 360–387 (2008)
5. Carter, C., Jennings, M.: Social responsibility and supply chain relationships. Transp. Res. E-Log. **38**(1), 37–52 (2002)
6. Murphy, P., Poist, R.: Socially responsible logistics: an exploratory study. Transp. J. **41**(4), 22–35 (2002)
7. Seuring, S.: Core issues in sustainable supply chain management—a Delphi study. Bus. Strateg. Environ. **17**(8), 455–466 (2008)
8. Elkington, J.: Cannibals with Forks: the Triple Bottom Line of 21st-Century Business. New Society Publishers, Gabriola Island, BC (1998)
9. Loza-Aguirre, E., Segura, M., Roa, H., Montenegro, C.: Unveiling unbalance on sustainable supply chain research: did we forget something? In: Rocha, Á., Guarda, T. (eds.) Proceedings of the International Conference on Information Technology and Systems (ICITS 2018). Advances in Intelligent Systems and Computing, vol. 721. Springer, Cham (2018)
10. Montenegro, C., Loza-Aguirre, E., Segura-Morales, M.: Using probabilistic topic models to study orientation of sustainable supply chain research. In: Rocha, Á., et al. (eds.) WorldCIST'18 2018, AISC, vol. 745. Springer, Cham (2018)
11. Muñoz, M., Rivera, J., Moneva, J.: Evaluating sustainability in organizations with a fuzzy logic approach. Ind. Manag. Data Syst. **108**(6), 829–841 (2008)
12. Vimal, K., Vinodh, S.: Development of checklist for evaluating sustainability characteristics of manufacturing processes. Int. J. Proc. Manag. Bench. **3**(2), 213–232 (2013)
13. Sloan, T.: Measuring the sustainability of global supply chains: current practices and future Directions. J. Glob. Bus Manag. **6**(1), 1–16 (2010)
14. Blei, D.: Probabilistic topic models. Commun. ACM **55**(4), 77–84 (2012)
15. Guyon, I., Elisseeff, A.: An introduction to variable and feature selection. J. Mach. Learn. Res. **3**, 1157–1182 (2003)
16. Chandrashekar, G., Sahin, F.: A survey on feature selection methods. Comput. Electr. Eng. **40**, 16–28 (2014)
17. Saeys, Y., Inza, I., Larrañaga, P.: A review of feature selection techniques in bioinformatics. Bioinform. Rev. **23**(19), 2507–2517 (2007)
18. Dy, J., Brodley, C.: Feature selection for unsupervised learning. J. Mach. Learn. Res. **5**, 845–889 (2004)
19. Law, M., Figueiredo, M., Jain, A.: Simultaneous feature selection and clustering using mixture models. IEEE Trans. Pattern Anal. Mach. Intell. **26**(9), 1154–1166 (2004)
20. Kohav, R., John, G.: Wrappers for feature subset selection. Artif. Intell. **97**, 273–324 (1997)
21. Bolón-Canedo, V., Sánchez-Maroño, N., Alonso-Betanzos, A.: Feature Selection for High-Dimensional Data. Springer International Publishing, Cham (2015)
22. MathWorks: Neural Network Toolbox™. User's Guide. R2014a, The MathWorks, Inc. (2014)
23. Barreto, G., Mota, J., Souza, L., Frota, R., Aguayo, L. Yamamoto, J., Macedo, P.: Competitive neural networks for fault detection and diagnosis in 3G cellular systems. In: de Souza, J.N., et al. (eds.) ICT 2004, Berlin (2004)
24. Kaur, R., Sachdeva, M., Kumar, G.: Study and comparison of feature selection approaches for intrusion detection. Int. J. Comput. Appl. (2016)
25. Arguello, B.: A Survey of Feature Selection Methods: Algorithms and Software, Austin (2015)
26. Friedman, N., Geiger, D., Goldszmidt, M.: Bayesian network classifiers. Mach. Learn. **29**, 131–163 (1997)

27. Cheng, J., Greiner, R.: Comparing Bayesian network classifiers. In: Proceedings of the Fifteenth Conference on Uncertainty in Artificial Intelligence, Stockholm (1999)
28. Parameter estimation for text analysis. http://www.arbylon.net/publications/text-est.pdf
29. Select number of topics for LDA Model. https://cran.rproject.org/web/packages/ldatuning/vignettes/topics.html
30. Griffiths, T., Steyvers, M., Tanenbaum, J.: Topics in semantic representation. Psychol. Rev. **114**(2), 211–244 (2007)
31. Deveaud, R., Sanjuan, E., Bellot, P.: Accurate and effective latent concept for ad hoc information retrieval. Rev. Sci. Tech. Inf. **17**, 61–84 (2014)
32. Arun, R., Suresh, V., Veni, C., Murthy, M.: On finding the natural number of topics with latent Dirichlet allocation: some observations. In: Zaki, M., Xu, J. (eds.) Advances in Knowledge Discovery and Data Mining, pp. 391–402. Springer, Heidelberg (2010)
33. Cao, J., Xia, T., Li, J., Zhang, Y., Tang, S.: A density-based method for adaptive LDA model selection. Neurocomputing **72**(7–9), 1775–1781 (2009)
34. Liu, L., Tang, L., Dong, W., Yao, S., Zhou, W.: An overview of topic modeling and its current applications in bioinformatics. SpringerPlus, vol. 5, nº 1608, pp. 1–22 (2016)
35. The University of Waikato, WEKA Manual for Version 3-7-8, Hamilton, New Zealand (2013)
36. Witten, I., Eibe, F., Hall, M.: Data Mining: Practical Machine Learning Tools and Techniques. The Morgan Kaufmann Series in Data Management Systems (2011)

# A Systematic Literature Review on Word Embeddings

Luis Gutiérrez and Brian Keith[(⊠)]

Department of Computing and Systems Engineering,
Universidad Católica del Norte, Av. Angamos 0610, Antofagasta, Chile
{luis.gutierrez,brian.keith}@ucn.cl

**Abstract.** This article presents a systematic literature review on word embeddings within the field of natural language processing and text processing. A search and classification of 140 articles on proposals of word embeddings or their application was carried out from three different sources. Word embeddings have been widely adopted with satisfactory results in natural language processing tasks in general and other domains with good results. In this paper, we report the hegemony of word embeddings based on neural models over those generated by matrix factorization (i.e., variants of word2vec). Finally, despite the good performance of word embeddings, some drawbacks and their respective solution proposals are identified, such as the lack of interpretability of the real values that make up the embedded vectors.

**Keywords:** Bayesian networks · Sentiment analysis · Literature review
Opinion mining

## 1 Introduction

This work is contextualized within the development of a thesis in the field of sentiment analysis, also known as opinion mining. This field focuses on the task of detecting, extracting, and classifying opinions, sentiments, and attitudes concerning different topics in a text. Sentiment analysis has a series of applications, such as the study of political movements in social networks, customer satisfaction, market intelligence, among others [1]. In this context, it is an interest of this systematic review to research vector representations of words to be inserted in semantic vector spaces, with the purpose of providing a formal support for the selection of representations in subsequent works of sentiment analysis. However, it should be noted that the results obtained in this review are also applicable to other fields that lie under the umbrella of natural language processing.

In some studies, the dimensionality of the sparse word-context matrix is reduced through techniques such as singular value decomposition (SVD) [2]. Recently, proposals have been made to represent words through dense vectors that are derived from various training methods, inspired by the modeling of languages through neural networks. These types of representations are referred to as "neural embeddings" or "word embeddings" [2].

© Springer Nature Switzerland AG 2019
J. Mejia et al. (Eds.): CIMPS 2018, AISC 865, pp. 132–141, 2019.
https://doi.org/10.1007/978-3-030-01171-0_12

Word embeddings have proven to be a mechanism with which the task of computing similarities between words is facilitated by performing efficient calculations through operations with low-dimensional matrices. On the other hand, they are efficient to train, highly scalable for large corpora (thousands of millions of words), and for vocabularies and contexts of similar proportions [2]. The justification for a systematic review lies in the need to summarize or synthesize existing information on a particular topic in an unbiased way [3] and replicable by future interested researchers. In this case, the topic of interest is the application of dense vector representations for words, previously mentioned as word embeddings. The elaboration of a systematic review on this topic becomes relevant, and also necessary, since dense representations of words in semantic vector spaces have played an important role for basic tasks of natural language processing [4, 5], as well as for more complex and pertinent tasks to this work, as is sentiment analysis [6, 7]. On the other hand, no previous systematic reviews were found whose main topic is word embeddings, therefore, this systematic review could be the starting point for subsequent studies.

## 2  Methodology

The systematic literature review was conducted according to the method proposed in [3], which involves the planning, execution and reporting stages of the research.

### 2.1  Research Planning

Research planning considers the following elements:

1. **Research question:** The fundamental research question of this systematic review is: What semantic vector space representations for words have been proposed and in what contexts?
2. **Keywords:** During the development of the investigation, a set of keywords in English, listed in Table 1 together with an approximate translation into Spanish were considered, since some of them do not have a direct translation. In addition, it is indicated if the plural form of the keyword in English was also considered for the investigation.

**Table 1.**  Keywords and their translation.

| English | Spanish | Plural |
|---|---|---|
| Word embedding | Incrustamiento de palabra | Yes |
| Word representation | Representación de palabra | Yes |
| Vector space | Espacio vectorial | Yes |
| Semantic | Semántico | No |

3. **Selection of sources:** As source selection criteria, the sources used were those whose search engines for articles, accessible within the bibliographic databases of

the University, provided scientific articles entirely free of charge, and not only their abstracts and/or bibliographical references. Specifically, the following sources of articles were considered: ACM Digital Library, Scopus, and Web of Science. Most of the articles present in these sources are in English; however, articles in Spanish have also been considered in the application of the selection criteria.

4. **Query string:** A search string was built in order to enter a query into the search engines available in the selected sources of articles, containing all the keywords mentioned above. The search string is as follows: ("word embeddings" OR "word embedding" OR "word representation" OR "word representations") AND ("vector space" OR "vector spaces") AND "semantic".

In addition to searches in the article sources according to the query string, a non-systematic search was also carried out considering bibliographic references of a set of articles, and opportunistic searches on the Internet, as proposed in [8], whose main topic are the representations of words in semantic vector spaces.

## 2.2 Study Selection Criteria

Once the initial search has been carried out, the study selection criteria are aimed at identifying those primary studies that provide evidence to answer the research question [3] of this work. Primary studies are understood as studies that contribute to systematic reviews; in contrast, systematic reviews themselves constitute secondary studies. In order to reduce possible biases that may arise, the selection criteria must be defined during the definition of the research protocol [3], and not after planning. The selection criteria of the primary studies are divided into Inclusion Criteria (CI), and Exclusion Criteria (CE). Some criteria, both Inclusion and Exclusion, were based and adapted from [9]. Figure 1 shows the flow chart with the sequential application of the inclusion criteria that were proposed, as well as their interaction with the exclusion criteria.

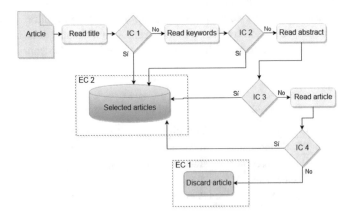

**Fig. 1.** Flowchart for selecting primary studies.

The Inclusion Criteria (IC) for each article were applied in the sequential order detailed below.

- IC 1: Articles whose title maintains a relationship with some or all the keywords established in this document will be included.
- IC 2: Articles whose keywords are a subset of the keywords established in this document will be included.
- IC 3: Those articles whose abstracts have descriptions or references that deal with representations of words in semantic vector spaces will be included.
- IC 4: Articles that present proposals on new models that generate representations of words in semantic vector spaces will be included.

The Exclusion Criteria (EC) are the following:

- EC 1: All articles that do not comply with any Inclusion Criteria, applied sequentially, will be excluded.
- EC 2: All articles that have already been reviewed in other sources will be excluded.

## 3   Results and Discussion

### 3.1   Results of the Review

As a first step in the execution of the systematic review, we proceeded to consult the sources of articles selected for the study with the search string constructed and previously presented. The results are shown in the second column of Table 2. The proportion of articles by source, together with the corresponding percentages, is also illustrated in Fig. 2(a).

**Table 2.**  Number of found and selected articles by source.

| Source | Number of found articles | Number of selected articles |
|---|---|---|
| ACM Digital Library | 14 | 7 |
| Scopus | 125 | 34 |
| Web of Science | 11 | 4 |
| Total | 150 | 45 |

After reading and applying both the inclusion and exclusion criteria, the articles pertinent to the purpose of this work were identified. The number of selected articles is shown in the third column of Table 2 and the proportion is shown graphically in Fig. 2(b).

### 3.2   Discussion

The 45 articles selected in this systematic review position word embeddings as a ubiquitous technique in sentiment analysis, information retrieval, and natural language processing in general. The relevant articles were defined as those that either had a new

**PROPORTION OF FOUND ARTICLES**

■ ACM Digital Library   ■ Scopus   ■ Web of Science

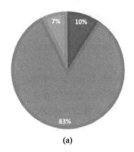

(a)

**PROPORTION OF SELECTED ARTICLES**

■ ACM Digital Library   ■ Scopus   ■ Web of Science

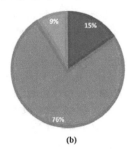

(b)

**Fig. 2.** (a) Graph of number of publications found by source. (b) Graph of number of publications selected by source.

theoretical proposal to generate word embeddings or those articles that included the concept of word embeddings in disciplines that were not usually directly related with natural language processing and the usage of word embeddings in conjunction with machine learning techniques. The use that the articles made of the concept of word embeddings are mentioned below, beginning with the two representations of words found in almost all the articles, used directly for the realization of the tasks or as a baseline for the evaluation of new proposals of word representations.

word2vec is the name by which are known two language models based on neural networks proposed by Mikolov et al. [10] to generate dense vector representations of words. Specifically, two model architectures are proposed: Continuous Bag of Words Model (CBOW), and Continuous Skip Gram Model (SG). On the one hand, CBOW aims to predict the occurrence of a word given other words that constitute its context. The context of a word $w_i$ is understood as the vicinity composed by the $k$ words to the left of $w_i$, and the $k$ words to the right of $w_i$. On the other hand, SG deals with predicting a context given the word $w_i$. $k$ is a hyperparameter of the model, known as the size of the local context window. In both cases, the models provide dense vector representations for the words that have proved effective in preserving the semantic characteristics of these despite a drastic reduction in dimensionality and training in a shallow neural network, which also speeds up the training process. Following the structure of this document, the works that are related to word2vec are not cited, because as mentioned above, they are almost all, which shows the success of these models.

The second vector representation that appears more recurrently in articles is the so-called Global Vectors for Word Representation (GloVe), proposed by Pennington et al. [11]. Unlike word2vec, which is a predictive model, GloVe is closer to a model that reduces the dimensionality of a co-occurrence matrix of the word-word type, generated by a fixed dimension local context window. GloVe gets its name from the fact that the statistics of the entire corpus (at a global level) are captured directly by the model. In addition, it proved to be competitive and reported better results than other state-of-the-art methods, such as word2vec, in tasks such as word analogy, word similarity, and Named Entity Recognition.

One of the main uses reported for word embedding is to perform semantic similarity estimation between words of different languages, which in essence goes back to the first applications of natural language processing. In this sense, in [12] a model is proposed to jointly learn bilingual embeddings based only on comparable data constituted by aligned documents that are in two different languages; this model was called Bilingual Word Embeddings Skip Gram (BWESG), which induces a multilingual vector space to embed word representations, queries, and even complete documents. In this same line, Glavaš et al. [13] proposes another method to measure textual semantic similarity between documents written in different languages, characterized by being light on resource usage, and that consists of linearly transferring representations of words from a vector space in a language of origin to a vector space in the target language. The word embeddings used in this work are generated using GloVe and CBOW.

There are also proposals for word embeddings for languages that have very particular alphabets, such as Arabic. Soliman et al. [14] propose a pre-trained set of models of word representations in the Arabic language, in order to provide the community with word embeddings generated from different domains, such as tweets, websites and Wikipedia articles in Arab. The disambiguation of words, a recurring task in the processing of natural language, has also been proposed to solve using word embeddings in Arabic; specifically, Laatar et al. [15] propose this solution in order to develop a dictionary that shows the evolution of the meaning and use of Arabic words, which in turn would help safeguard the Arab cultural heritage. It should be noted that these articles are based on word embeddings generated by the word2vec architectures.

Another common use of word embeddings is reported, which consists of their incorporation into recommender systems. In this sense, Musto et al. [16] present a preliminary investigation for the adoption of word embeddings in which both objects and user profiles are embedded in a vector space, to be used by a content-based recommender system. Boratto et al. [17] state that the use of word embeddings in content-based recommender systems is less effective than other collaborative strategies (for example, the decomposition of singular values), so it proposes another approach, defining a vector space in which the similarity between an object that has not been evaluated by the user and those objects that have been evaluated is measured in terms of linear independence, reaching better results than, for example, SVD. Greenstein et al. [18] assert to be able to convert to words the sequence of objects sought by a user and thus be able to project them in a vector space, in such a way that similarity and analogies between objects can be detected. The word embeddings of these articles are generated according to the word2vec and GloVe models.

The use of word embeddings is also reported in conjunction with other machine learning techniques or linguistic resources. Alsuhaibani et al. [19] establish that the methods that generate vector representations of words based purely on information distributed in a corpus, fail to take advantage of the semantic relational structure that there is between words in concurrent contexts; These structures are detailed in manually elaborated knowledge bases, such as ontologies and semantic lexicons, where the meaning of words is defined by the various relationships that exist between them. Therefore, the corpus is combined with the knowledge bases to generate word embeddings that, when used, present an improvement in performance in the tasks of measuring similarity and analogy of words, obtaining results that support the hypothesis.

On the other hand, Liu [20] proposes that, in addition to generating vector representations of words having the corpus as a source, internal elements of each word, such as morphemes, should also be taken into consideration. For this, two models are proposed to generate word embeddings: Morpheme on Original view and Morpheme on Context view (MOMC) and Morpheme on Context view (MC), which show higher performance in detecting the similarity of words than the models of the baseline, among which was CBOW. Gallo et al. [21] present a method in which the word embeddings generated by word2vec are encoded in images, to later make use of convolutional neural networks (CNN) and perform text classification on the images. The method reported better classification results when compared to the baseline (doc2vec with SVM).

Wild and Stahl [22] present an implementation of Latent Semantic Analysis and its results when generating word representations in vector spaces. Unlike word2vec and GloVe, this method is based on the factorization of matrices, due to the use of singular value decomposition.

One of the main drawbacks of word embeddings is the decrease in the dimensionality of the problem at the expense of the interpretability of the real values that make up the vector representations, which is commonly known as an opaque model. Given this, solutions have been proposed to make the representations interpretable. Among them, Liu et al. [23] propose techniques for visualizing useful semantic and syntactic analogies in various domains, using as base representations those generated by word2vec and GloVe. Andrews [24] points out that the representations learned by the word embeddings models, despite having small dimensions, still make use of an important amount of storage, so the use of Lloyd's algorithm is proposed since it can compress dense representations by a factor of 10 without further penalizing performance. In addition, an efficient factorization method of computing in GPU is presented to obtain representations with greater interpretability, each dimension is coded with a non-negative value; on the other hand, they are also sparse. The tasks of similarity and analogy of words with the compressed representations were evaluated and it was demonstrated that the aspirations of the work were attainable.

Regarding the relevant articles found in opportunistic searches, Joulin et al. [25] present a method to generate word embeddings several orders of magnitude faster than the models resulting through deep learning. The method is similar to CBOW, replacing the word from the context medium with a classification label, obtaining a performance on par with models that take more time to train in tasks such as sentiment analysis and label prediction. Moody [26] presents a method in which the Skip Gram architecture of word2vec is mixed with the modeling of topics in documents that make use of the Latent Dirichlet Allocation technique. The model is called lda2vec and is capable of generating vector representations of words and documents in the same vector space. Next, Table 3 shows the most important articles on the investigation, considering proposed models that have been widely adopted in the literature to generate word embeddings; reported improvements to these models; the incorporation of these with other natural language processing techniques (e.g. Latent Dirichlet Allocation); and also, possible solutions to the problem of interpretability that these dense vector representations pose. It must be noted that these criteria answer to the necessity of highlighting articles that are appropriate to future studies, according to the context of the authors' work, and not necessarily to the most frequently cited articles.

**Table 3.** Number of found and selected articles by source.

| Year | Title | Authors |
|------|-------|---------|
| 2007 | Investigating Unstructured Texts with Latent Semantic Analysis | Fridolin Wild, Christina Stahl |
| 2013 | Efficient estimation of representations in vector space | Tomas Mikolov, Kai Chen, Greg Corrado, Jeffrey Dean |
| 2014 | GloVe: Global Vectors for Word Representation | Jeffrey Pennington, Richard Socher, Christopher D. Manning |
| 2016 | Bag of Tricks for Efficient Text Classification | Armand Joulin, Edouard Grave, Piotr Bojanowski, Tomas Mikolov |
| 2016 | Mixing Dirichlet Topic Models and Word Embeddings to make lda2vec | Christopher Moody |
| 2016 | Compressing Word Embeddings | Martin Andrews |

# 4   Conclusions

Dense vector representations of words have been widely adopted with satisfactory results in natural language processing tasks in general. In addition, they have been applied in other domains with good results. It is also worth mentioning one of the limitations of this study, namely that the work has been done on only three databases of scientific publications (ACM, Scopus and Web of Science).

On the other hand, the hegemony of word embeddings based on neural models over those generated by matrix factorization is reported; numerous alternatives have been proposed that are variants of a group of neural models: word2vec. In this context, studies have been conducted on the impact of including other techniques together with word embeddings to perform natural language processing tasks, reporting good results.

As future work, applying the inclusion and exclusion criteria, and the systematic review process used in this work to other scientific publications databases could be a starting point in order to extend the scope of this review. This review could be used as the foundation for further analysis of the literature in the medium and long term, considering the volume of research that is done on word embeddings in different areas.

Finally, despite the good performance of word embeddings, some drawbacks and their respective solution proposals are identified, such as the lack of interpretability of the real values that make up the embedded vectors.

**Acknowledgments.** Research partially funded by the National Commission of Scientific and Technological Research (CONICYT) and the Ministry of Education of the Government of Chile. Project REDI170607: "Multidimensional Bayesian classifiers for the interpretation of text and video emotions".

# References

1. Ravi, K., Ravi, V.: A survey on opinion mining and sentiment analysis: tasks, approaches and applications. Knowl. Based Syst. **89**, 14–46 (2015)
2. Levy, O., Goldberg, Y.: Dependency-based word embeddings. In: Proceedings of the 52nd Annual Meeting of the Association for Computational Linguistics (Volume 2: Short Papers), vol. 2, pp. 302–308 (2014)
3. Kitchenham, B.: Procedures for performing systematic reviews. Keele, UK, Keele University, 33(2004), 1–26 (2004)
4. Collobert, R., Weston, J.: A unified architecture for natural language processing: deep neural networks with multitask learning. In: Proceedings of the 25th International Conference on Machine Learning, pp. 160–167. ACM (2008)
5. Zou, W.Y., Socher, R., Cer, D., Manning, C.D.: Bilingual word embeddings for phrase-based machine translation. In: Proceedings of the 2013 Conference on Empirical Methods in Natural Language Processing, pp. 1393–1398 (2013)
6. Severyn, A., Moschitti, A.: Twitter sentiment analysis with deep convolutional neural networks. In: Proceedings of the 38th International ACM SIGIR Conference on Research and Development in Information Retrieval, pp. 959–962. ACM (2015)
7. Tang, D., Wei, F., Yang, N., Zhou, M., Liu, T., Qin, B.: Learning sentiment-specific word embedding for twitter sentiment classification. In: Proceedings of the 52nd Annual Meeting of the Association for Computational Linguistics (Volume 1: Long Papers), vol. 1, pp. 1555–1565 (2014)
8. Carrizo Moreno, D.: Atributos contextuales influyentes en el proceso de educción de requisitos: una exhaustiva revisión de literatura Ingeniare. Revista Chilena de ingeniería **23** (2), 208–218 (2015)
9. Díaz, N.D., Zepeda, V.V.: Ejecución de una Revisión Sistemática sobre Gestión de Calidad para Sistemas Multiagente
10. Mikolov, T., Chen, K., Corrado, G., Dean, J.: Efficient estimation of word representations in vector space. arXiv preprint arXiv:1301.3781 (2013)
11. Pennington, J., Socher, R., Manning, C.: Glove: global vectors for word representation. In: Proceedings of the 2014 Conference on Empirical Methods in Natural Language Processing (EMNLP), pp. 1532–1543 (2014)
12. Vulíc, I., Moens, M.F.: Monolingual and cross-lingual information retrieval models based on (bilingual) word embeddings. In: Proceedings of the 38th International ACM SIGIR Conference on Research and Development in Information Retrieval, pp. 363–372. ACM (2015)
13. Glavaš, G., Franco-Salvador, M., Ponzetto, S.P., Rosso, P.: A resource-light method for cross-lingual semantic textual similarity. Knowl. Based Syst. **143**, 1–9 (2018)
14. Soliman, A.B., Eissa, K., El-Beltagy, S.R.: AraVec: a set of Arabic word embedding models for use in Arabic NLP. Procedia Comput. Sci. **117**, 256–265 (2017)
15. Laatar, R., Aloulou, C., Bilguith, L.H.: Word sense disambiguation of Arabic language with word embeddings as part of the creation of a historical dictionary. In: Language Processing and Knowledge Management Proceedings. CEUR-WS.org (2017)
16. Musto, C., Semeraro, G., de Gemmis, M., Lops, P.: Learning word embeddings from Wikipedia for content-based recommender systems. In: European Conference on Information Retrieval, pp. 729–734. Springer (2016)
17. Boratto, L., Carta, S., Fenu, G., Saia, R.: Representing items as word-embedding vectors and generating recommendations by measuring their linear independence. In: RecSys Posters (2016)

18. Greenstein-Messica, A., Rokach, L., Friedman, M.: Session-based recommendations using item embedding. In: Proceedings of the 22nd International Conference on Intelligent User Interfaces, pp. 629–633. ACM (2017)
19. Alsuhaibani, M., Bollegala, D., Maehara, T., Kawarabayashi, K.I.: Jointly learning word embeddings using a corpus and a knowledge base. PloS One **13**(3), e0193094 (2018)
20. Liu, J.: Morpheme-enhanced spectral word embedding. In: Proceedings of the International Conference on Software Engineering and Knowledge Engineering (2017)
21. Gallo, I., Nawaz, S., Calefati, A.: Semantic text encoding for text classification using convolutional neural networks. In: 2017 14th IAPR International Conference on Document Analysis and Recognition (ICDAR), vol. 5, pp. 16–21. IEEE (2017)
22. Wild, F., Stahl, C.: Investigating unstructured texts with latent semantic analysis. In: Advances in Data Analysis, pp. 383–390. Springer (2007)
23. Liu, S., Bremer, P.T., Thiagarajan, J.J., Srikumar, V., Wang, B., Livnat, Y., Pascucci, V.: Visual exploration of semantic relationships in neural word embeddings. IEEE Trans. Visual Comput. Graph. **24**(1), 553–562 (2018)
24. Andrews, M.: Compressing word embeddings. In: International Conference on Neural Information Processing, pp. 413–422. Springer (2016)
25. Joulin, A., Grave, E., Bojanowski, P., Mikolov, T.: Bag of tricks for efficient text classification. arXiv preprint arXiv:1607.01759 (2016)
26. Moody, C.E.: Mixing Dirichlet topic models and word embeddings to make lda2vec. arXiv preprint arXiv:1605.02019 (2016)

# Model for the Improvement of Knowledge Management Processes Based on the Use of Gamification Principles in Companies in the Software Sector

Jose Luis Jurado[1]([✉]), Diego Fernando Garces[1],
Luis Merchan Paredes[1], Emilia Rocio Segovia[1],
and Francisco Javier Alavarez[2]

[1] Carrera 122 #6-65, Cali, Colombia
{jljurado,dfgarces,lmercahn,rsegovia}@usbcali.edu.co
[2] Av. Universidad #940, Ciudad Universitaria, Aguascalientes, Mexico
fjalvar.uaa@gmail.com

**Abstract.** The Knowledge management is a fundamental and complex process for nowadays organizations. There are several factors that affect its implementation, in practice; the low motivation of people belonging to organizations imposes an interesting challenge to multiple research trends and areas. The approach addressed in this paper to encourage and facilitate people engaged in such process is termed gamification. Gamification is based on games principles. It is useful for solving engagement issues within organizations applied in processes. This paper introduces an adapted model to the conditions of software organizations allowing identification, creation, application, transferring and evaluation of knowledge, using gamification principles. Model application and obtained results are discussed and analyzed, allowing its repeatability by the reader. It was concluded that results of the presented research-in-progress are promising.

**Keywords:** Knowledge management · Gamification
Software process improvement

## 1 Introduction

The improvement of knowledge management processes is the main motivation of this work, these processes require effort to perform tasks or activities such as: create and build data, transform information and later generate the knowledge, so says [1]. This implies executing practices such as: identify data, select resources, group information, make associations, document processes, assign responsibilities, define involved, among many others according to [2].

The purpose of this paper is to motivate the analysis of alternatives to support the difficulties presented in organizations when they implement knowledge management processes. In particular, those associated with complexity in the understanding and subsequent execution of processes in knowledge management.

© Springer Nature Switzerland AG 2019
J. Mejia et al. (Eds.): CIMPS 2018, AISC 865, pp. 142–151, 2019.
https://doi.org/10.1007/978-3-030-01171-0_13

One of the difficulties identified is that, related to the large number of existing models, initiatives and methodologies that, due to their diversity and focus, make it difficult for small and medium enterprises to implement procedures and formats that are not aligned and adapted to the real conditions of these organizations, so he says [3].

Other authors such as [4] have identified circumstances associated with low motivation as: lack of interest on teamwork, inefficient tools in knowledge management processes; inadequate controls in the processes of identification, transformation and knowledge transferring; in addition an inappropriate allocation of tasks and roles; rules of interaction that do not exist among the members of an organization, among many others.

To complement the above, is highlighted a study presented in [5] conducted by a research group in Colombia with the support of a group of companies (25) in the software sector in the city of Cali (Colombia), the objective was to make a diagnosis in tools and good practices in knowledge management.

The result of the investigation allowed to identify factors associated to the failure in the development of knowledge management processes such as: disinterest in sharing knowledge; lack of incentives and methods to encourage participation and inclusion in knowledge management processes; ignorance of common practices and documentation regarding the subject in context. In this sense and considering works such as [6], it has been determined that a common element found in different scenarios and contexts of knowledge management and its applications is motivation.

The motivation can be analyzed from two points of view: The first, when there is a high motivation; in this case the work group shows high participation, integration and interest rates in the knowledge management processes. And second, when there is low motivation; in this case the work group experiences a low production of knowledge artifacts and low involvement in activities related to the acquisition and transfer of knowledge, as described by works such as [7].

The document initially presents the proposed model framed in a systematic review that gives it a theoretical basis. Later, a case study is described that allowed to validate the logic and functionality of the model, using a Web tool as an instrument to facilitate the implementation of the case study. Finally, the conclusions and future work for this project are presented, which has other validations planned in different companies of the software sector.

## 2 Knowledge Management Model Adapted

The methodology for designing the model proposed in this document is based on a literature review. Initially, the contributions of different authors in knowledge management models and perspectives are considered. The most relevant ideas are then integrated into an initial design, which structures its logical and procedural model, based on the type of functionalist and interpretive knowledge management, as mentioned by [8]. This work has been taken as a reference to establish the types of knowledge management, considering the degree of maturity of an organization, from the perspective of its knowledge management activities.

The proposed structure for the model considers two main components. The Fig. 1 shows the conceptual vision of the model proposed, integrated by two components: A set of knowledge management processes and the application context, adapted version of works such as [9], but with a flexible vision of the knowledge generation and conservation processes, adapted to the conditions of software organizations.

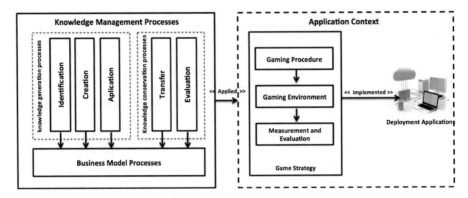

**Fig. 1.** Knowledge management model

The *knowledge management processes* component consists of 3 subcomponents (knowledge generation processes, knowledge conservation processes and business model processes). The idea of grouping common tasks such as (Identify, create, apply, transfer and evaluate) in two sub-processes (generation and conservation) facilitate flexibility in the execution of organization's business model processes. This perspective has been considered from works such as [10], where they propose the subdivision of tasks and roles in knowledge management processes.

The proposal to define common knowledge management activities has been taken from work such as [11], where tasks and documents associated with the identification, creation, application, transfer and evaluation of knowledge have been identified.

This perspective is also considered in works like [10], where they associate common tasks in an organization with the knowledge transformation cycle. Another initiative that has been considered is the proposal in [9], where they associate knowledge generation and conservation activities with the processes of conversion from tacit to explicit knowledge.

The other component that is described in Fig. 1 is *the application context*, which is in the lower part of the proposed model; in this component the game strategy that will be implemented in the business model is defined, it is important to highlight the work of [12], which was taken as a reference, where it's proposed a method for the design of game strategies that can be applied in any organizational context.

This method is composed by three phases (a game procedure, an environment of game and the measurement and adjustment of the strategy), the reason for using gamification principles in organizational processes has been discussed in works such as [9], which raises the need to streamline knowledge management processes for the organizations.

The reason for using gamification principles to improve knowledge management processes is because it is intended to positively impact the interactions between stakeholders and the knowledge generation and conservation processes.

## 3   Study Case

This section describes a implementation of a case study, based on references such as [13], that has been considered, to validate the effectiveness of the proposed model. In this sense, initially, a context of the organization is presented where the case study is to be developed; later the objectives and scope of the validation are proposed, next the way to apply the proposed model for the case study is described, finally, are present the results of the validation and an analysis of the lessons learned.

### 3.1   Context and Purpose of Study Case

The company selected is a medium-sized company in the software industry, provider of information technology development services, focused on the architecture, design and implementation of Java EE, BPM and SOA projects. Its business model is the development of software service providers in online-health sector oriented products. The selected product for the study case is a web services and mobile client platform. The product development time is 4 months, and the evaluated modules are registration and care of incidents.

The purpose of the study case is to verify the impact of the use of game strategies in knowledge generation processes, the subprocess conservation will not be analyzed in this validation, since initially in the research that is being developed it presents a first phase in relation to the activities related to the identification, creation and application of knowledge. In a later phase of the project it is expected to analyze aspects related to the transfer and application of knowledge, and the relationship of the use of game strategies in the improvement of the production of the final product of the organization.

The validation that has been defined for the study case is related to motivation, and its influence on knowledge generation. The indicators established are participation and contribution of explicit knowledge.

These indicators have been defined considering works such as [14], where he proposes the analysis of the participation in software development activities. Other work mentioned in [15] describes the relationship of the quality of the contribution in a software product, with high motivation in her production. Therefore, the following metrics are proposed to estimate the impact of motivation in the sub-process of knowledge generation.

*Participation in the identification and creation of explicit knowledge*: A member of the software development team has the ability to share his best practices, for the development of software code packages or software design artifacts.

*Contribution in the application of knowledge*: The member of the development team feeds back the knowledge shared, adding good practices of their learned lessons, during the application of knowledge shared in design and implementation artifacts and software tests.

## 3.2    Implementing the Study Case Improvement Model

The implementation of the model proposed in this paper proposes to perform two tasks: Establishment of knowledge management processes improvement and design of the game strategy. These tasks are aligned with the two components defined in the proposed model. The first is related to the "knowledge management processes" component, and the second one with the component of "Application context".

*Establishment of knowledge management processes improvement* defines a set of activities associated with the process of knowledge conservation. Which can be grouped into the sub-processes of identification, creation and application of knowledge. Table 1, shows a characterization of the activities identified in the organization, during software development process and their relationship with the knowledge generation sub processes, and with the areas of the organization involved in said processes.

**Table 1.**  Activities associated with the knowledge generation process.

| Conservation process | Associated activities | Areas of the organization involved |
|---|---|---|
| Identification | Identification of basic knowledge units | Development team |
|  | Identification of existing and non-existent areas of knowledge | Requirement team |
|  | Identification of profiles that favor the creation of new knowledge | |
| Creation | Refinement of basic units of knowledge | Requirement team |
|  | Creation of tacit knowledge to explicit knowledge | Development team |
|  | Knowledge integration | |
| Application | Use in solving problems and projects | Team solutions and business architects |
|  | Use in organizational learning | |
|  | Use for organizational strategies | |

The previous table shows an example of how to establish activities associated with a business model, with knowledge management processes.

Once established the relationships between activities, sub processes and their impact on the areas of the organization in the business model, it's necessary to define the procedures into the organization, that allow those involved to have a guide for establishing their participation in the different activities of knowledge generation, for reasons of industrial secrecy said procedures can not be published.

*Design of the game strategy*, the second task "game strategies design" allows to establish the way in which knowledge management activities are executed from the perspective of gamification. This task integrated in three phases (Gaming Procedure, Gaming Environment and Measurement and Evaluation) has been defined considering different works that propose processes and procedures to elaborate a game strategy. In particular, the work of [16] has been considered, where he proposes a method called DeDalus as a tool for the design of a game strategy. The strategy proposed for the study case using this method is described below in a summarized form.

*Gaming Procedure*: The design of the game is aimed at rewarding the fulfillment of explicit objectives in a software development process, from its elicitation of requirements, the design of its architecture, the construction of its source code and its subsequent validation through test cases.

The game strategy proposes to classify the performance of players through medals (apprentice, master and senior): The (apprentice) is the player who is in the process of creating knowledge applying practices of their peers to produce learned lessons. This distinction shows how a player has the ability to transform explicit knowledge in tacit knowledge, through artifacts such as "published practices". The (master) is the player who has an interest in applying knowledge and actively participate in the creation and identification of software development practices. He has the ability to contribute to the knowledge created by their peers, this can be evidenced through the number of "used practices" that they have registered. Finally, the (senior) is the player that refines knowledge, through the artifact "refined practices".

*Gaming Environment*: The game environment describes the rules and restrictions that the game establishes for the proper application of the strategy. For the study case that is being analyzed, the Table 2 shows an example of some actions during the game and their respective score. It is important to clarify that you can define more game actions and also associate them with specific activities of the sub processes. But for purposes of the study case only some were taken as an example.

**Table 2.** Game rules

| Task development | Artifact | Sub process |
|---|---|---|
| Register a practice | Published practice | Creation |
| Record a learned lesson from the use of a published practice | Practice used | Identification |
| Propose an improvement in the application of a practice | Refine practice | Application |

The scores can vary and be adjusted according to the results obtained once the strategy is validated, the DeDalus method is designed to measure and adjust the rules of the game according to the performance of the strategy

*Measurement and Evaluation*: During this stage it's suggested to define a way to evaluate the effectiveness of the game strategy, establishing indicators that allow analyzing the impact degree of the game elements, as it considers it [17]. For this purpose, two indicators are defined (participation and contribution) to measure the impact of motivation in knowledge generation activities. Each indicator has three possible values (high, medium, low). These are assigned according to the achievement (apprentice, master and senior) obtained during the game.

Finally, the deployment of improvement strategy it was carried out through a Web application. This tool has three basic services: User registration and definition of the player profile; the use of a collaborative environment where good practices are uploaded and interact with the contributions of the work team. Likewise, the application allows visualizing the scores in a leaderboard and obtaining badges that identify the achievement in the knowledge generation of the players.

### 3.3    Preliminary Results

This section shows the partial results, in relation to the artifacts generated in each of the activities of the sub processes of knowledge generation. Table 3 shows a summary of the results of the development team and its production, during the 4 months that was taken as a sample to collect data.

**Table 3.** Results of a pilot test

| Development team | Published practices | Used practices | Practice contribution | Achieved insignia | Related subprocess |
|---|---|---|---|---|---|
| Player 1 | 13 | 10 | 10 | Senior | Application |
| Player 2 | 10 | 8 | | Master | Identification |
| Player 3 | 4 | | | Apprentice | Creation |
| Player 4 | 12 | 8 | 6 | Senior | Application |
| Player 5 | 8 | 6 | | Master | Identification |
| Player 6 | 10 | 9 | | Master | Identification |
| Player 7 | 9 | 8 | 5 | Senior | Application |
| Player 8 | 12 | 5 | 7 | Master | Identification |
| Player 9 | 8 | | | Apprentice | Creation |
| Player 10 | 12 | 10 | 6 | Master | Identification |

The table above shows a significant number of artifacts generated in the different sub-processes of knowledge generation. It was possible to motivate 3 engineers to reach the senior category, which shows their high interest not only in producing knowledge, but also in the use and improve of what you learn, enriching the said practices with learned lessons.

50% of totally of participants reached the master category, this represents that the knowledge generated in the development team has a good acceptance and approval and that generated practices are effectively used in the development processes of each member.

Finally, it should be noted that it was possible that 100% of those involved were carried away by the game strategy. Achieving only 2 amateurs category and generated 98 knowledge practices, figures that confirm that the motivation was positively affected in relationship with figures obtained from similar projects, where the activity of producing knowledge of the development team was very scarce.

The evaluation indicators (participation and contribution) reflect that in relation to the participation of those involved in knowledge generation activities, a high measurement was achieved. The results show that 5 badges of master were obtained, that is to say, 64 artifacts of used practices were produced. This leads to analyze that the proposed strategy allows to positively inducing those involved, by generating an active participation in the production of knowledge.

The contribution indicator allows to highlight that the motivation has a positive effect, when refining or improving the practices that have been used by the work team. This enrichment is possible by recording learned lessons, in the practices to work team.

For this measurement, 5 senior badges were obtained, which implies a production of 34 refined practices, obtaining a high contribution by the development team, this data is very important since it allows analyzing the impact of motivation when knowledge is codified or recorded correctly, allowing an adequate interpretation of it and its subsequent execution.

In relation to the results of knowledge management activities, Table 4 shows the artifacts generated and the details of each activity carried out by the stakeholders.

**Table 4.** Results of artifacts generated in knowledge management activities

| Task Development | Generated artifact | Quantity | Format | Language | Sub processes |
|---|---|---|---|---|---|
| Register a Practice | Work schedule | 1 | PMI | Spanish | Identification |
| | Negotiation proceedings | 1 | Proposed by the company | Spanish | |
| | Source code | 15 | Scripts | Java, html | Creation |
| | Test plans | 4 | IEEE 829 | Spanish | Application |
| Record a lesson learned | Requirements document | 3 | IEEE 829 | Spanish | Creation |
| | Source code | 11 | Scripts | Angular, SQL | |
| | Architecture Design | 3 | ISO/ISE/IEEE 42010 | C4, 4 + 1 | Application |
| Application of a practice | Source code | 2 | Scripts | SQL | Application |
| | Test plans | 2 | IEEE 829 | Spanish | |

In the case of the practices of a development process, positively influencing the contribution of knowledge promotes teamwork and the improvement of their development practices, since many software products require the refinement and feedback of different actors in a software development process.

Table 4 shows that the knowledge application sub-process has been present in most of the activities carried out by those involved. And the "Register a practice" activity has the most artifacts generated. The artifact with the most interactions is the "source code" artifact, especially during knowledge creation. In addition, the "source code" artifact is the most recurrent in knowledge management activities and sub-processes.

## 4   Conclusions

The figures obtained in the study case are inconclusive, in relation to the impact of motivation and its relationship with knowledge management processes, which are designed from the perspective of gamification, but allow to contribute to a line of study, in relation with proposals or alternatives to improve the production of knowledge, in processes as complex as software development, and in particular to generate knowledge from the transformation of explicit knowledge to tacit.

The problematic exposed in this work, was focused on analyzing the impact of motivation in the generation of knowledge in organizations that need to implement formal processes of knowledge management. Particularly the drawbacks related to motivation were related to the comprehension of knowledge management activities, their adaptation to the organizational context and their subsequent application. These disadvantages were addressed from the perspective of proposing a model adapted to the contexts of software organizations, this perspective allowed to offer a model that proposes an initial stage of characterization of common activities of knowledge management in frequent processes such as the generation of knowledge.

The benefit of grouping common activities in knowledge management processes makes it possible to control these activities individually and to estimate their impact on the organization. This also allows adaptations to specific contexts in a specific business models, for the case of software development contexts, its adaptation was positive since it was beneficial grouped specific activities of knowledge generation, allowing a process of knowledge transformation from tacit to explicit in the generation of knowledge artifacts, produced in the different stages of the software development process.

The published practices allowed to evidence unidentified knowledge in requirements elicitation activities, design of software architectures, good practices in the use of programming languages and test cases. In addition, this adaptation allowed the organization to discover new knowledge through the learned lessons in the use of software development practices, this knowledge will be used and transmitted throughout the organization to improve their organizational processes and enrich the quality of their products

The use of gamification techniques for the design of game strategies has been positive in its purpose of impacting participation and contribution in the production of knowledge. This does not mean that it is the only and best alternative, many others that are being implemented in organizations to improve the problems related to formalism and the adaptation of knowledge management processes are valid and very accurate. In fact, the gamification takes as reference techniques and methods from areas such as psychology, economics, and social sciences among others.

## References

1. Aurum, A., Daneshgar, F., Ward, J.: Investigating Knowledge Management practices in software development organisations - An Australian experience (2014)
2. Bjørnson, F.O., Dingsøyr, T.: Knowledge management in software engineering: a systematic review of studied concepts, findings and research methods (2014)
3. De Vasconcelos, J.B., Kimble, C., Carreteiro, P., Rocha, Á.: The application of knowledge management to software evolution. Int. J. Inf. Manag. 37(1), 1499–1506 (2017)
4. Sharp, H., Baddoo, N., Beecham, S., Hall, T., Robinson, H.: Models of motivation in software engineering. Inf. Softw. Technol. 51(1), 219–233 (2009)

5. Morales, J.D., Jurado, J.L.: Caracterización en la gestión y adopción de herramientas y estrategias de gestión de conocimiento del sector de TI del Valle del Cauca, trabajo de titulo de grado. Universidad de San Buenaventura, Colombia (2016)
6. Rus, I., Lindvall, M., Sinha, S.S.: Knowledge management in software engineering: a DACS state-of-the-art report. Fraunhofer Center for Experimental Software Engineering Maryland and The University of Maryland, 19, 26–38 (2011)
7. Dalkir, K., Beaulieu, M.: Knowledge Management in Theory and Practice. MIT press, Cambridge (2017)
8. González, C.A.R., Muñoz, M.E.Q., Yepes, C.M.D.: Metodología para evaluar la madurez de la gestión del conocimiento en algunas grandes empresas colombianas. Tecnura: Tecnol. Cult. Afirmando El Conoc. 19(43), 20–36 (2015)
9. Jurado, J.L., Collazos, C.A., Gutierrez, F.L., Paredes, L.M.: Designing game strategies: an analysis from knowledge management in software development contexts. In: International Conference on Serious Games, Interaction and Simulation, pp. 64–73. Springer, Cham (2016)
10. Galvis-Lista, E., Sánchez-Torres, J.M.: A critical review of knowledge management in software process reference models. JISTEM-J. Inf. Syst. Technol. Manag. 10(2), 323–338 (2015)
11. Nonaka, I., Krogh, G.V., Voelpel, S.: Organizational knowledge creation theory: evolutionary paths and future advances. Organ. Stud. 27(8), 1179–1208 (2006)
12. Aparicio, A., Montes, J., Vela, F.: Analysis of the effectiveness of gamification. In: APPLEPIES Proceedings, pp. 108–119 (2016)
13. Runeson, P., Höst, M.: Guidelines for conducting and reporting case study research in software engineering. Empir. Softw. Eng. 14(2), 131 (2009)
14. Hamri J., Jonna K., Harri S.: Does gamification work? — A literature review of empirical studies on gamification. In: Proceedings of 47th Hawaii International Conference on System Sciences, USA (2014)
15. Werbach, K.: (Re) defining gamification: a process approach. In: International Conference on Persuasive Technology, pp. 266–272. Springer, Cham, May 2014
16. Jurado, J., Andrea Boxiga, P., Brigitte González, L., Felipe Bacca, A.: Using DeDalus to design a game strategy to improve motivation in learning processes of basic programming, pp. 8854–8864 (2017). https://doi.org/10.21125/iceri.2017.2455
17. Seaborn, K., Fels, D.I.: Gamification in theory and action: a survey. Int. J. Hum. Comput. Stud. 74, 14–31 (2015)

# Intelligent Mathematical Modelling Agent for Supporting Decision-Making at Industry 4.0

Edrisi Muñoz[1]([⊠]) and Elisabet Capón-García[2]

[1] Centro de Investigación en Matemáticas A.C.,
Jalisco S/N, Mineral y Valenciana, 36240 Guanajuato, Mexico
emunoz@cimat.mx
[2] ABB Switzerland Ltd., Segelhofstrasse 1K, 5405 Baden-Dättwil, Switzerland
elisabet.capon@ch.abb.com

**Abstract.** The basis of decision-making at industry consists of formally representing the system and its subsystems in a model, which adequately captures those features that are necessary to reach consistent decisions. New trends in semantics and knowledge models aim to formalize the mathematical domain and mathematical models in order to provide bases for machine reasoning and artificial intelligence. Hence, tools for improving information sharing and communication have proved to be highly promising to support the integration of performance assessment within industrial decision-making. This work presents an intelligent agent based on knowledge models and establishes the basis for automating the design, management, programming and solution of mathematical models used in the industry. A case study concerning a capacity limitation constraint demonstrates the performance of the agent and indicates the directions for future work.

**Keywords:** Process optimization · Knowledge management · Intelligent agent
Mathematical modelling · Mathematical programming

## 1 Introduction

Nowadays, industry is deep into in the fourth industrial revolution, the so-called Industry 4.0, where computers and work automation interact together. In this context, sensors and actuators should be connected remotely to computer systems, equipped with sophisticated algorithms that can understand, learn and decide with small intervention from human operators.

Basically, the basis of those interactions is in charge of decision support systems (DSSs). The success on improving decision-making depends on how analytical models and data are integrated. On the one hand, the correct modelling task, when a problem or certain reality is found, increases the success of the decision. On the other hand, the responsiveness and accuracy to get information quality and smart data complement how good the decision can be.

J. Mejia et al. (Eds.): CIMPS 2018, AISC 865, pp. 152–162, 2019.
https://doi.org/10.1007/978-3-030-01171-0_14

Thus, models can be classified as qualitative and quantitative according to the manner in which information is represented. Qualitative models usually represent the physical and logic relationships among the elements of the system describing the reality, such as conceptual or semantic models. In contrast, quantitative models allow proposing decisions based on the analysis of actual data regarding the system, such as, mathematical or statistical models. Thus, mathematical modelling has a key role in many fields such as the environment and industry, while its potential contribution in many other areas is becoming more and more frequently.

This work particularly focuses on providing and enhancing the support of mathematics used for describing and representing objects and devices behavior from the system within the structure of mathematical models.

The goal of this work consists of creating an intelligent agent based on knowledge models capable of unfolding the complexity of mathematical models. Specifically, given a mathematical equation describing the relations within a physical or organizational system, the intelligent agent should read the equation, identify its components and relations, collect knowledge to conceptualize these components and relations, and formalize everything in a computer and human understandable language. Therefore, this agent stands for a key piece in a wider framework for designing, managing, programming and solving mathematical models used within the process industry.

This contribution is structured as follows. First, a brief introduction to Industry 4.0 is provided. Next, the methodology describes the mathematical modelling and the agent algorithm. Finally, a case study is presented in order to show how the intelligent agent performs. Specifically, an equation related to capacity limitation in a scheduling model is used to demonstrate how the framework performs tasks related to: (i) exploitation of the equation, (ii) concept relation, (iii) mathematical programming translation, and (iv) the structuring and management of data required by mathematical programming. Finally, conclusions are discussed and future work is proposed.

## 1.1   Industry 4.0

Along time, industry has suffered important technological changes that have transformed and improved human life. First, during the late seventeenth and the early eighteenth centuries, machines and steam came to mechanize some industrial work tasks. Next, during the late eighteenth and earliest twentieth centuries, electrification and production lines allowed the birth of mass production. A third era came with the advent of computers, the beginnings of automation and specifically with the internet, here robots and machines began to replace human workers on assembly lines. Finally, the fourth revolution, the so-called Industry 4.0, has arrived by introducing the "smart factories", where cyber-physical systems monitor the physical processes of the factory and make decentralized decisions. Here, the physical systems aim to communicating and cooperating both with each other and with humans in real time.

The characteristics of this type of industry comprise the implementation of features such as: decentralized decision making, integrated transactional system environment, synchronized real and virtual systems, and implemented decision support system.

- Decentralized decision making refers to the ability of the system elements for making decisions to become as autonomous as possible, focused on general goals.
- Integrated transactional systems aim to connect people, machines, sensors, devices and third parties by means of quality data coming from material requirements planning (MRPs), distribution resource planning (DRPs), and enterprise resource planning (ERPs) systems.
- Synchronized real and virtual systems refer to the capability of contextualizing the real system in a virtual one in order to allow problems exploration and experimentation without affecting the real one.
- Implemented decision support system aims to technically assist and support humans in making decision and solving problems.

Additional features can be listed for supporting this industry 4.0, such as, data security, optimum system reliability and system stability. Besides, according to experts [3, 4] the Internet of things (IoT), cyber-physical systems and big data, must help to achieve this big transformation of the industry. Finally it is important to remark that industry is concerned with the loss of high-paying human jobs when new automations are introduced, as well as, the potential of causing expensive production due to technical problems in any of the different actors of the production process.

### 1.2   Models for Decision Making

In industrial processes, modelling is of paramount importance as the basis for decision making task. The need for robust and flexible models able to cover the various physical and chemical aspects and diverse temporal and special scales involved on the processes is more evident. This implies being able to pose complex partial differential algebraic models, in pure equation form.

Moreover, models must be capable of combining quantitative and qualitative models through equations and logic. Finally, the capability of automating problem formulation, modelling, programming and solutions through higher level and detail of process features can have an enormous impact for the industry.

## 2   Explicit Concept-Object Oriented Mathematical Modelling

The proposed work is based on knowledge management models, working as a core reasoning motor for automated mathematical modelling. The approach focuses on formalizing the knowledge of aspects and mathematical operations contained in mathematical models. The goal of this proposal consists of developing an intelligent agent supported by explicitly linking mathematical elements, expressions and equations of mathematical models with their corresponding real conceptualization in a system, represented by domain semantic models. Finally, the intelligent agent aims to design, understand, manage and exploit of mathematical models. The following subsections will explain the main building blocks of the intelligent agent previously defined.

## 2.1    Ontological Models

Semantic models seem to be a promising technology for knowledge capturing and formalization. Besides, ontologies can also serve as tool for systematic standardization and homogenization of data and information found within transactional systems. In this work, two main ontologies, namely mathematical modelling ontology and enterprise ontology project, serve as the basis for the development of the intelligent agent.

### 2.1.1    Mathematical Modelling Ontology

The first semantic model consists of the Mathematical Modelling Ontology (MMO) [1, 2], which is in charge of modelling the mathematical domain knowledge. This knowledge model aims to semantically represent mathematical models of any engineering domain. MMO formally captures the main elements for mathematical modelling and formalizes the relation of these elements to the engineering domain concepts they represent. Therefore, MMO basically consists of the following components:

(i)    Mathematical elements, namely terms, logic expressions and algebraic expressions; as well as mathematical logic and algebraic operations
(ii)   Mathematical element behavior properties, such as variables, binary variable, constants, etc.
(iii)  Object-oriented relationships among mathematical elements and a semantic representation of the system they represent.

As a result, MMO enables the direct association of all the elements found in a mathematical model with the corresponding engineering concepts, which are unified/standardized semantically to classes, properties, and axioms of an existing domain ontology. Therefore, although the modeler is still responsible for verifying such links, this framework provides a tool for semantically formalizing them, thus clarifying and unveiling the assumptions in the modelling process of any mathematical model.

In order to recognize and parse the mathematical equations into MMO, it has been decided to adopt MathML as the language to write the equations. Figure 1 shows an UML scheme of the main classes and properties that formalize the vocabulary and relations for translating general mathematical equations into this ontology.

### 2.1.2    Enterprise Ontology Project

The second semantic model stands for the Enterprise Ontology Project (EOP), which encompasses the representation of industrial process domain [5]. This model aims to formalize the concepts related to industrial behavior and most of the actors found there comprising several functions, such as production, marketing, sales, human resources, logistics, safety and environment. This ontology describes four main areas, specifically process, procedure, recipe and physical areas, that combined allow many industrial functions. This approach follows the ANSI/ISA standards 88 and 95, used for control and automation of functions in process industries [6–9].

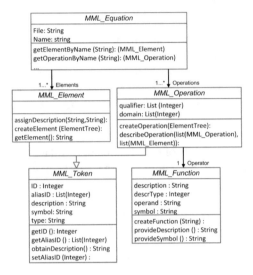

**Fig. 1.** UML representation of the main classes for automated parsing and recognition of mathematical equations.

## 2.2 Intelligent Agent

The aim of this research is the development and application of an intelligent agent to support the design, management, programming and solution of mathematical models used in the industry for enterprise-wide integration. Besides, this agent will support the reduction of the gap between conceptual and qualitative models; engineering-machine communication; and mathematical model reusability and adaptation.

Therefore, the agent coordinates the functions of parsing the mathematical model, connecting it to the process semantic model, saving and retrieving existing models, proposing existing solution methods, and communicating with decision makers. A simplified overview of the agent functions is provided in Fig. 2. The intelligent agent has been developed in Jython programming language [10]. Jython is a Java implementation of Python that combines a more flexible environment to Java. Thus, while working in Java, Jython libraries can be added allowing to write simple or complicated scripts that add functionality to the application.

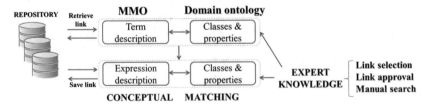

**Fig. 2.** Agent functions (represented as rows) related to interaction between ontological system representation, mathematical model repository and expert knowledge.

The architecture of this agent is based on model-view-control approach (MVC). MVC is a standard software architecture that separates data from an application, the user interface, and control logics, into three distinct components called layers [11]. Besides, this architecture allows the scaling of the infrastructure easily. The main features of the layers are explained next.

**Model Layer.** It is also called business layer (or business logic) because all the rules that must be accomplished are set here. In general, a program that is running in an application first receives a user request, then it sends responses after some process has been executed. This layer communicates with the view layer, to receive inputs and to send outputs that are presented as results. Finally, this layer communicates with the control layer in order to ask the database manager to store or retrieve data from it.

**View Layer.** This layer aims to interact with the final users by delivering data and information on desire formats. The view represents the page design code, and the Controller represents the navigational code.

**Control Layer.** This layer is also known as data and access layer, being in charge of accessing and managing data. It consists of one or more database administrators to perform all data storage, receiving requests for storing or retrieving information from the business layer.

In this work, the model layer contains the main functions (algorithm) of the designed agent. Then, view layer facilitates the interaction with engineers, when human-machine and/or machine-machine interactions occur. Finally, the control layer consists of the ontological domain models and related functions for reasoning and getting requested data and information.

## 2.3 Agent Algorithm

This section presents the main algorithm for automation of the explicit concept-object oriented mathematical modelling. It is important to remark that the next requirements must be taken in to account in order to use the intelligent agent.

- An abstraction of a physical system to be studied. It refers to a problem to be trade must exist.
- A mathematical model(s) of a problem concerning the system also must be available. The mathematical model must be in "MathML content" format [12]. Besides a file with nomenclature in "cvs" format must exists.
- A database for mathematical models comprising different problems, realities and physical and chemical phenomenal, must be identified. Once the agent has introduced a mathematical model or problem to the system, all the knowledge is stored and reused for future decisions.
- A domain ontology on web ontology language (OWL) language should be available for concept-object knowledge extraction. The system can be expanded as many as semantic models are introduced in the system.

The main algorithm consists of four main phases, namely (a) Expression identification; (b) Term pattern recognition, (c) Conceptual connector; and (d) Expression pattern recognition, as shown in Fig. 3.

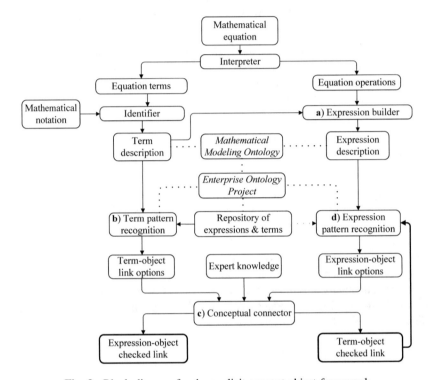

**Fig. 3.** Block diagram for the explicit concept-object framework.

- **Expression builder.** In this phase "mathematical operations" are analyzed in order to define expression sequencing. The Expression Builder constructs a natural language description of a certain expression based on: (i) the matching of the operation itself with the content MathML class, and (ii) the description of the dependent "Term description" and "Expression description". The inputs of this phase, coming from requirement previously established are a MathML equation and a nomenclature file. As a result, the natural language term and expression description are used for term and expression pattern recognition.

- **Term pattern Recognition.** In this phase semantic resemblance association is done. Thus, all the possible associations between specific terms and existing objects from a domain ontology is developed. Specifically, this function searches for semantic resemblance between the term description (nomenclature) and classes, data and object properties found in the domain Ontology. Additionally, a "Repository of terms and expressions" is used in order to find similar expressions. The inputs of this phase are a nomenclature file and information from the domain ontology (concepts and properties).

- **Conceptual connector.** The "Conceptual connector" aims to collect expert knowledge and assign the "Term" and "Expression" to the adequate object in the domain ontology. This block allows: (i) all the "Term-object" links to be related to their corresponding objects, and subsequently the "Expression pattern recognition" is applied, and (ii) "Expression-object" links to be established according to the described procedure. Inputs in this phase are basically expert decision (for starting use of the agent), and the output results in a complete concept-to-object definition for a given equation.
- **Expression pattern recognition.** This phase analyses the semantic description of the expressions derived from the "Expression builder". Next, based on the "Term-object" checked link, the agent unveils all possible "Expression-object" link options. Besides, a repository allows to trace back equivalences among concepts in the expressions. This provides a list of objects from a domain ontology, related to the "Expression" in semantic and conceptual likeness.

## 3    Case Study

The functions of the intelligent agent are illustrated using a capacity limitation constraint in the context of modelling process scheduling problems. Therefore, the capacity limitation constraint is given as defined in Eq. (1):

$$S_{s,t} \leq C_s^{max} \; \forall s, t \tag{1}$$

where the notation "symbol-type-description" is established as follows: $s$-set-state; $t$-set-time period; $C_s^{max}$-parameter-maximum storage capacity; $S$-variable-available amount.

The mathematical agent parses and recognizes the building elements of the previous equation and based on a notation list, it automatically generates the list of elements and expressions as presented in Fig. 4. Likewise, not only does the agent

**MML_Element**

| ID | Symbol | Alias | Type | Value | Description |
|----|--------|-------|------|-------|-------------|
| 1 | S | [1] | variable | | storage level |
| 2 | s | [2, 5, 6] | set | {s1..s10} | state |
| 3 | t | [3, 7] | set | {t1..t10} | time |
| 4 | Cmax | [4] | parameter | | maximum storage capacity |

Loaded equation: Capacity_limitation

**MML_Operation**

| ID | Element | Alias | Operand | Domain | Description | Symbol |
|----|---------|-------|---------|--------|-------------|--------|
| 8 | expression | [8] | exprseq | [9, 12] | storage level in t... | ((S_{st} <= Cma... |
| 9 | expression | [9] | leq | [10, 11] | storage level in t... | (S_{st} <= Cmax... |
| 10 | variable | [10] | selector | [1, 2, 3] | storage level in t... | S_{st} |
| 11 | parameter | [11] | selector | [4, 5] | maximum stora... | Cmax_{s} |
| 12 | expression | [12] | forall | [13] | for all state and ... | (forall (s and t)) |
| 13 | expression | [13] | and | [6, 7] | state and time | (s and t) |

**Fig. 4.** Automatically generated list of MML elements (top) and MML expressions (bottom) for the capacity limitation constraint.

reproduce the capacity limitation constraint, but also creates a natural language description of the equation terms and expressions according to Table 1.

**Table 1.** Description in natural language automatically generated for the MML expressions of the capacity limitation constraint.

| ID | Description | Symbol |
|----|-------------|--------|
| 8 | storage level in the set of state and time is lower than or equal to maximum storage capacity in the set of state for all state and time | S_{st} <= Cmax_{s}) (for all (s and t)) |
| 9 | storage level in the set of state and time is lower than or equal to maximum storage capacity in the set of state | S_{st} <= Cmax_{s} |
| 10 | storage level in the set of state and time | S_{st} |
| 11 | maximum storage capacity in the set of state | Cmax_{s} |
| 12 | for all state and time | forall (s and t) |

These initial tasks are aligned to allow the understanding of the proposed model for capacity limitation by means of the processing of the mathematical equation. Hence, not only the equation is exploited and translated into a machine understandable model, but also the symbol-concept relation is managed by the intelligent agent. Therefore, Fig. 5 shows the screens provided by the framework and related to the linking between mathematical and domain representations.

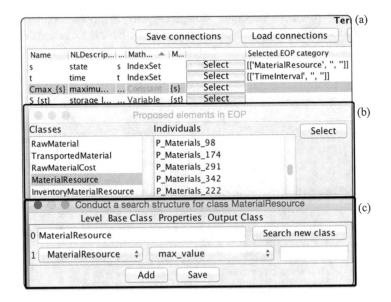

**Fig. 5.** (a) List of MMO terms related to MML terms and expressions, defining the index sets, parameters and variables. (b) Suggested links between $C_s^{max}$ and the domain ontology (EOP). (c) Dialog for class and property selection.

Finally, after completing the connection between the mathematical model and the semantic representation of the domain; the agent is capable of connecting to the databases in order to provide the required data to solve the mathematical programming associated with the whole set of equations for a given problem.

## 4   Conclusions

This work presents an intelligent agent which allows for the automatic parsing, interpretation and formalization of mathematical equations in a human and computer understandable structure. This agent relies on previously developed semantic models and is based on a "concept-object" framework developed in Python for directly interpreting mathematical equations of a given problem formulation. In addition, the decomposition of the equation into its mathematical elements and the linking to domain concepts allow to easily verify and validate the model and enables direct human interaction and machine reasoning. Thus, mathematical models related to different enterprise processes can be validated based on the same domain concepts, facilitating their integration among different time and space scales and hierarchical levels of decision-making. Therefore, this contribution stands for a cornerstone for further developments in the integration of mathematical models in industrial practice using artificial intelligence and machine learning methods. Specifically, a framework for automatically designing, managing, programming and solving mathematical models has been devised for the process industry, and will be presented in future work.

## References

1. Munoz, E., Capon-Garcia, E., Lainez, J., Espuna, A., Puigjaner, L.: Integration of enterprise levels based on an ontological framework. Chem. Eng. Res. Des. **91**, 1542–1556 (2013)
2. Bender, E.: An Introduction to Mathematical Modelling. Dover Publications, New York (1978)
3. Gubán, A., Kása, R.: Conceptualization of fluid flows of logistificated processes. Adv. Logistic Syst. **7**(2), 27–34 (2013)
4. Mayer, V., Kenneth Cukier, S.: Big Data, A Revolution That Will Transform How We Live, Work, and Think (2013)
5. Munoz, E., Capon-Garcia, E., Espuna, A., Puigjaner, L.: Ontological framework for enterprise-wide integrated decision-making at operational level. Comput. Chem. Eng. **42**, 217–234 (2012)
6. International Society for Measurement and Control. Batch control part 1 models and terminology (1995)
7. International Society for Measurement and Control. Control batch part 4 batch production records (2006)
8. International Society for Measurement and Control. Batch control part 5 automated equipment control models & terminology (2007)
9. International Society for Measurement and Control. ISA-88/95 technical report: using isa-88 and isa-95 together. Technical report (2007)

10. Marzal Varó, A., García Sevilla, P., Gracia Luengo, I.: Introducción a la programación con Python 3. Castellón de la Plana: Universitat Jaume I. Servei de Comunicació i Publicacions (2014)
11. Avgeriou, P., Zdun, U.: Architectural patterns revisited: a pattern language. In 10th European Conference on Pattern Languages of Programs, pp. 1–39 (Euro-Plop 2005). Irsee (2005)
12. Mathematical markup language (MathML) version 3.0. http://www.w3.org/TR/2010/REC-MathML3-20101021

# Software Systems, Applications and Tools

# MBT4J: Automating the Model-Based Testing Process for Java Applications

Leonardo Villalobos-Arias[(⊠)], Christian Quesada-López,
Alexandra Martinez, and Marcelo Jenkins

University of Costa Rica, San Pedro, Costa Rica
{leonardo.villalobosarias, cristian.quesadalopez,
alexandra.martinez, marcelo.jenkins}@ucr.ac.cr

**Abstract.** Model-based testing is a process that can reduce the cost of software testing by automating the design and generation of test cases but it usually involves some time-consuming manual steps. Current model-based testing tools automate the generation of test cases, but offer limited support for the model creation and test execution stages. In this paper we present MBT4J, a platform that automates most of the model-based testing process for Java applications, by integrating several existing tools and techniques. It automates the model building, test case generation, and test execution stages of the process. First, a model is extracted from the source code, then an adapter—between this model and the software under test—is generated and finally, test cases are generated and executed. We performed an evaluation of our platform with 12 configurations using an existing Java application from a public repository. Empirical results show that MBT4J is able to generate up to 2,438 test cases, detect up to 289 defects, and achieve a code coverage ranging between 72% and 84%. In the future, we plan to expand our evaluation to include more software applications and perform error seeding in order to be able to analyze the false positive and negative rates of our platform. Improving the automation of oracles is another vein for future research.

**Keywords:** MBT4J platform · Model-based testing · Software testing
Model extraction · Test generation · Test execution · Adapter · Automation

## 1 Introduction

Model-based testing (MBT) refers to the automatic derivation of tests from a model of the software under test (SUT) [1]. MBT techniques can provide high quality test suites by systematically generating test cases until the desired coverage criterion is reached [2]. MBT has been shown to find a larger number of defects than manual testing [3, 4]. Moreover, MBT has been reported to reduce the overall cost, time, and effort of the testing process [3, 5].

Industry adoption for model-based testing has been slow [1] and is still in an introductory stage [6]. While MBT offers several benefits over manual testing, it requires a greater initial effort investment [7]. Particularly, the creation of the model and the infrastructure for test case execution have been reported as the most effort-consuming

© Springer Nature Switzerland AG 2019
J. Mejia et al. (Eds.): CIMPS 2018, AISC 865, pp. 165–174, 2019.
https://doi.org/10.1007/978-3-030-01171-0_15

tasks [7–10]. Moreover, model building lacks a systematic methodology and supporting tools for its automation [11]. To address these problems, the automation of both model creation [1] and test adapters [12] has been proposed as a possible solution. Automating part of the testing process can improve its effectiveness and significantly reduce development costs [13] thus increasing the feasibility of its adoption by the industry [11].

We present here MBT4J, a platform built with the goal to automate the MBT process in the context of Java applications, by leveraging existing tools and techniques such as model and adapter generation. From the SUT's source code, our platform can create a model, generate tests cases and execute them on the SUT. We conducted a case study to assess the efficacy of MBT4J. The chosen SUT was a Java calculator from a public repository. Efficacy was measured in terms of both model and SUT coverage, number of defects, and number of test cases.

The paper is structured as follows: Sect. 2 contains an overview of the model-based testing process. Section 3 shows previous works on MBT tools and approaches. Section 4 explains the MBT4J platform. Section 5 describes the evaluation performed on the platform. Section 6 outlines our conclusions and future work.

## 2　The Traditional Model-Based Testing Process

The traditional model-based testing process comprises four stages: (1) building the model, (2) choosing the test selection criteria, (3) generating the test cases and (4) executing them [1]. Next we describe of these stages in detail.

The **model building stage** encompasses the creation of a model from the requirements. This model represents the expected behavior of the system under test at an abstract level. The model design or creation is usually a manual task.

The **test selection criteria stage** guides the test generation process by choosing a finite but complete set of test cases—from the possibly infinite set of all test cases that can be derived from the model. Choosing the right test selection criteria depends on the type of SUT and test objectives.

The **test case generation stage** explores the model in an automated way, and produces a set of test cases that meets the test selection criteria. These test cases are abstract—which means that they are at the same level of abstraction as the model—and need to be *concretized* to be run on the SUT.

The **test case execution stage** involves concretizing the abstract test cases generated in the previous stage and executing them on the SUT—either manually or automatically. Two approaches exist: online and offline. In offline execution, test cases are generated before they are executed. In online execution, test cases are executed as they are generated.

## 3　Related Work

Although traditional model-based testing deals with the automation of test case generation, automation of the other stages has been implemented through tools. We present and group existing MBT tools according to their degree of automation.

Existing tools aid in the MBT process in varying degrees of automation. The most basic group of MBT tools are only able to generate abstract test cases from a model [14–16], relying on the tester to perform all other stages of the process. The second group of tools generates the tests and also automates (or aids in) their execution or concretization [8, 17, 18]. The third group of tools supports all MBT stages [2, 19, 20], but are particularly limited in their support for model and adapter creation, steps that still need to be performed manually by the tester. Our MBT4J platform would fit into the third group of tools, additionally automating the model and adapter generation.

Many studies have performed evaluations on MBT tools. Some studies aimed at evaluating a single tool with one case study [8, 9, 21]. Other studies have taken a comparative approach, either comparing a specific tool against others [13, 22], or comparing tools within a particular context [10]. These studies often reported challenges in the modeling stage and the construction of the test adapter.

Our study will advance the current state of the art by providing a platform that automates the model building, test case generation, and test case execution stages of the MBT process. We achieved this by expanding the capabilities of a tool from the third group, ModelJUnit. Thus, our platform not only supports but automates all MBT stages.

## 4    The Proposed Platform: MBT4J

MBT4J is the proposed platform for automating the model-based testing process of Java applications. Since it uses the online test execution approach, MBT4J merges the test generation and test execution stages of the MBT process into a single phase, effectively yielding a three-stage process. Hence, the process supported (and partially automated) by our platform consists of three stages (same description as in Sect. 2): (1) model building, (2) test selection criteria, and (3) test case generation and execution.

The MBT4J platform contains three main components: the model extraction component, the adapter generation component, and the test case generation and execution component. Figure 1 shows these components and their internal steps, with intermediate and final products. It additionally depicts the logical flow between components and internal steps. The model extraction component instruments the SUT to obtain context information, which is then transformed into a behavioral model (Sect. 4.1). The adapter generation component further expands the model's capabilities by adding an adapter that maps the abstract model actions into executable SUT calls (Sect. 4.2). The test case generation and execution component explores the model to simultaneously generate and execute the tests on the SUT, using the adapter (Sect. 4.3). The model extraction component belongs to the model building stage of the MBT process, whereas both the adapter generation component and the test case generation and execution component belong to the test generation and execution stage of the MBT process.

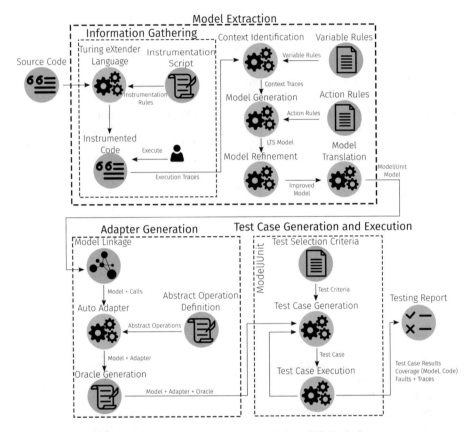

**Fig. 1.** Main components and flow of the MBT4J platform

## 4.1 Model Extraction

The objective of the model extraction component is to systematically and automatically generate a model from the SUT's source code. For this, we use a model extraction technique based on context information [23]. Each context is an abstract state of the program, composed of the execution point and the data state of the program [23]. The model extraction component has five steps.

The **information gathering** step takes the SUT's source code and obtains information about the system, in the form of execution traces. For this, the source code is passed through a systematic instrumentation process [23]. The resulting code will produce traces of the current context information when each execution point is reached. In order to obtain the execution traces, it is necessary to execute the instrumented code, exercising the behavior that one wants to model. We provide automation for this step in the form of a previously existing annotation script [24], with modifications, for the TXL language [25].

The **context identification** step takes the execution traces and identifies each context, assigning them a unique identifier. The result is a series of context traces, and a table that maps these traces to the original context information. This transformation is performed by our own implementation of an identification algorithm [23], which requires the list of variables that will be considered for context identification.

The **model generation** step creates a behavior model based on the identified context traces. This model represents the SUT behavior exercised during the information gathering step. We use our own implementation of the algorithm that transforms the context information into a behavior model [23], based on the list of actions that will considered for the creation of the model.

The **model refinement** step improves the generated model. Part of this process is performed by the LTSA tool [26], which runs a minimization algorithm on the model. The resulting model is then manually checked by the tester against the expected results. If the behavior is inconsistent, any of the previous steps is repeated with modifications to the instrumentation script, or the variable, or the action lists, until the generated model represents the expected behavior.

The **model translation** step takes the refined model and translates it into the notation that the test generation tool takes as input. In our case, we provide a tool that converts the model into the ModelJUnit notation.

## 4.2    Adapter Generation

The objective of the adapter generation component is to systematically generate an adapter, i.e., an interface that maps abstract model actions into executable SUT calls. Automatically generating the adapter can be beneficial, since its otherwise manual development is labor-intensive and error-prone [12].

The AutoAdapter extension for ModelJUnit [12] is a tool that partially automates the adapter creation process, and we use it as part of our platform. This tool is linked with the existing ModelJUnit model. The end result is the execution of concrete SUT operation calls as the model is explored by ModelJUnit. If the SUT call requires parameters, AutoAdapter can generate this data. The adapter generation component has three steps.

The **abstract operation definition** step is concerned with the creation of an operation mapping. This mapping relates an abstract model action with one or multiple SUT operations. It also indicates the parameters and return types for each operation. Data generators can be specified for all parameters. This mapping is implemented as a Java class, in a format that AutoAdapter can understand. The result of this step is an operation class that links abstract and executable operations.

The **model linkage** step is concerned with linking the model with the AutoAdapter tool. The result is a ModelJUnit model that can execute test cases on the SUT. This process is performed by modifying the model so that it calls the adapter in each of the actions of the model. The abstract operation definition is used in this step so that the adapter knows which methods to call when the action is invoked.

The **oracle creation** step is concerned with the specification of an oracle: a piece of code that asserts whether the action performed was successful or not. These oracles are to be incorporated manually in the actions of the model. This step is crucial, as it verifies that the SUT produces a valid result. The result of this step is a model that can execute the test cases and assert that the SUT is working as intended.

### 4.3    Test Generation and Execution

The objective of the test generation and execution component is to systematically and automatically generate and execute test cases, while verifying the correct SUT behavior. This is performed by the ModelJUnit tool, based on the model obtained from the previous component. The end result is a test report that includes the executed test cases, the achieved coverage, and the defects found. Three steps make up this component.

The **test criteria selection** step refers to the selection of the parameters supported by the test generation tool. ModelJUnit supports three parameters: the model, the tester, and the total length. The model is the SUT model obtained from previous components. The tester is the generation algorithm that the tool employs to explore the model. ModelJUnit currently employs five testers: Random, Lookahead, Greedy, All-Round, and Quick. Each tester has their particular configuration. The total length indicates the length, measured in transitions, of the generated test suite.

Coverage metrics and other listeners can be added to a tester. The test criteria selection process usually needs to be performed multiple times until the desired coverage is obtained. When all these parameters are set, ModelJUnit is ready to generate test cases.

The **test case generation** step involves exploring the model to create test cases that satisfy the previously defined criteria. The result of this step is a set of test cases together with the model coverage they achieve. This process is carried out entirely by ModelJUnit, and no further action is required from the tester until all test cases have been generated.

The **test case execution** step involves executing the test cases on the SUT. We use the online approach for test case execution, so test cases are executed as they are generated. The result of this step is a report of defects found in the SUT, and the action trace that led to those defects, and code coverage attained by these test cases.

## 5    Evaluation

### 5.1    Design

To evaluate the MBT4J platform, we conducted a small-scale, technology-oriented case study. The case study aimed to answer the following research question: *What is the efficacy of the MBT4J platform?* Efficacy was measured in terms of number of test cases, number of defects, and coverage of model and SUT.

The SUT used for the validation of our platform was a pocket calculator, chosen from a public repository of Java applications [27]. This application was given as input to our platform. A model of the SUT was automatically built by the *model extraction* component. Then, an adapter was produced by the *adapter generation* component. Finally, the *test case generation* and *execution* component was invoked several times, to obtain all combinations of levels under study. There were 2 factors: the ModeJUnit tester and the total length. We evaluated 4 levels for the ModeJUnit tester: Random, Lookahead, Greedy and Allround, and 3 levels for the total length: one hundred, one thousand, and ten thousand. All tasks in the testing process were conducted by the first author.

## 5.2   Results

We were able to successfully automate most of the time-consuming and complex stages of the MBT process for the chosen SUT, by using the MBT4J platform. The model extraction process was conducted in 1:46 min, the adapter generation in 3:34 min, and the test case generation and execution in 0:39 min. The duration of the test case and execution component depended on the test length parameter. Regarding the model complexity, MBT4J generated an intermediate model with 341 states and 375 transitions, which was then simplified to 2 states and 27 transitions by the automatic model refinement step.

Table 1 details each configuration of MBT4J platform: tester (Tester column) and the total length (Length column). Additionally, it shows their collected efficacy metrics: number of generated test cases (Tests column), number of defects found (Defects column), model coverage (Model Coverage columns) and SUT coverage (Code Coverage column). Model coverage is measured as: (i) the percent of transitions explored (Transition column), and (ii) the percent of pairs of adjacent transitions explored (Transition Pairs column). On the other hand, code coverage is measured as the percent of lines of code tested (LOC column).

**Table 1.** Efficacy metrics of the MBT4J platform.

| Test configuration | | Tests | Defects | Model coverage | | Code coverage |
|---|---|---|---|---|---|---|
| Tester | Length | | | Transition | Transition pair | |
| Random | 100 | 8 | 2 | 96.3% | 20.9% | 74.41% |
| Lookahead | 100 | 3 | 1 | 100.0% | 6.5% | 71.86% |
| Greedy | 100 | 12 | 1 | 100.0% | 19.3% | 73.56% |
| Allround | 100 | 23 | 2 | 100.0% | 16.0% | 75.91% |
| Random | 1,000 | 93 | 31 | 100.0% | 86.4% | 78.46% |
| Lookahead | 1,000 | 3 | 1 | 100.0% | 6.5% | 71.64% |
| Greedy | 1,000 | 82 | 28 | 100.0% | 88.6% | 81.45% |
| Allround | 1,000 | 243 | 17 | 100.0% | 83.7% | 81.24% |
| Random | 10,000 | 941 | 271 | 100.0% | 100.0% | 82.73% |
| Lookahead | 10,000 | 3 | 3 | 100.0% | 6.5% | 72.92% |
| Greedy | 10,000 | 885 | 885 | 100.0% | 100.0% | 82.94% |
| Allround | 10,000 | 2438 | 142 | 100.0% | 100.0% | 83.58% |

From Table 1, we observe that MBT4J was able to achieve high model and SUT coverage. Most ModelJUnit testers attained 100% transition coverage with a total length of 100, and 100% transition pair coverage with a total length of 10,000. Code coverage over 70% was obtained with a total length of just 100. Increasing the total length to 10,000 only increased the code coverage to 83%. Both Random and Greedy testers were able to find the highest number of defects (289), while retaining code coverage from 74% to 83%. The Allround tester was not as effective at finding defects (up to 142), but provided a slightly higher coverage in most cases, ranging from 76% up to 84%. The Lookahead tester showed no variation across different total length configurations, since it stops generating tests when all transitions have been explored.

Across all total length configurations, the Allround tester generated more but shorter tests, while the Random and Greedy testers generated fewer but longer tests. Thus, the latter were better suited for finding defects associated to long action sequences, while the former was better at finding defects near the initial application state. Since the random and greedy testers found the highest number of defects in our SUT, we can infer that many defects required long action sequences to be detected.

An analysis of all defects found by MBT4J reveals that most of them relate to two main bugs in the application: (1) it forgets a pending operation when other operations are performed, and (2) it does not properly handle a division by 0 in some cases. Other defects were found by performing specific action sequences.

### 5.3   Discussion

*Findings.* First, from a modeling perspective, analyzing the complexity and accuracy of the generated model is relevant. Our platform can automatically generate low complexity models. Complex SUT models are automatically simplified it in the model refinement step. Besides, the platform allows the user to validate the model by checking if it correctly represents the behavior of the SUT. Second, from a testing perspective, it is useful to analyze the relationship between number of tests, number of detected defects, and coverage. In our case study, it seems that the both Greedy and Random testers offer a good compromise between test suite size, number of defects found, and coverage. Third, from a parameter tuning perspective, it is also useful to analyze the trend in number of test cases, number of defects found, and coverage when increasing the total length parameter. Preliminary results from our study point to a linear trend, except for coverage. When increasing the total length by a factor of 10, a similar increase is observed in the number of test cases and the number of defects found, but not so in the coverage (with a rather negligible increase). This can be caused by either unreachable code in the SUT or an incomplete model.

*Limitations.* First, from the modeling perspective, creating the model's oracles is a time consuming task, since it still is a manual step which requires understanding of both model and expected behaviour of the SUT. Also, the following tasks are still manual: determining which variables and actions are relevant for the model, and exercising SUT's behaviour. We agree with Duarte et al. [23] in that the quality of the final model depends on the knowledge and customization done by the tester. Second, from the empirical validation point of view, the limitations of our study include: the selection of a single and simple SUT, and the collection of metrics by only one researcher.

## 6  Conclusions and Future Work

We presented MBT4J, a model-based testing platform that can automatically generate and execute test cases for Java applications. The platform automates model building, test case generation, and test execution stages. Our platform is able to extract a behavior model from the SUT and aids in the adapter creation task. Both of these activities are reported as the most complex in MBT.

Our case study validated the efficacy of the platform. MBT4J was able to generate up to 2,438 test cases, detect up to 289 defects, and achieve up to 83% code coverage, and 100% transition and transition pair coverage.

Future work encompasses further evaluating MBT4J with more complex Java systems in order to obtain additional evidence about the efficacy of the platform. Also, seeding errors in these systems to analyze the false positive and negative rates. Finally, we plan on improving the automation of oracles within our platform, as this step was the most time consuming.

## References

1. Utting, M., Legeard, B., Bouquet, F., Fourneret, E., Peureux, F., Vernotte, A.: Recent advances in model-based testing. In: Advances in Computers, vol. 101, pp. 53–120. Elsevier (2016)
2. Utting, M., Legeard, B.: Practical Model-Based Testing: A Tools Approach. Morgan Kaufmann, San Francisco (2010)
3. Pretschner, A., Prenninger, W., Wagner, S., Kühnel, C., Baumgartner, M., Sostawa, B., Zölch, R., Stauner, T.: One evaluation of model-based testing and its automation. In: Proceedings of the 27th International Conference on Software Engineering, pp. 392–401. ACM (2005)
4. Broy, M., Jonsson, B., Katoen, J.P., Leucker, M., Pretschner, A.: Model-based testing of reactive systems. In: LNCS, vol. 3472. Springer (2005)
5. Jard, C., Jeron, T.: TGV: theory, principles and algorithms. Int. J. Softw. Tools Technol. Transfer 7, 297–315 (2005)
6. SwissQ: Testing trends & benchmarks 2013. where do we stand where are we going to? (2013) https://swissq.it/wp-content/uploads/2016/02/Testing-Trends_und_Benchmarks2013.pdf
7. Schulze, C., Ganesan, D., Lindvall, M., Cleaveland, R., Goldman, D.: Assessing model-based testing: an empirical study conducted in industry. In: Proceedings of the 36th International Conference on Software Engineering. ACM (2014)
8. Ernits, J., Roo, R., Jacky, J., Veanes, M.: Model-based testing of web applications using NModel. In: Testing of Software Communication Systems. Springer (2009)

9. Pinheiro, A.C., Simao, A., Ambrosio, A.M.: FSM-based test case generation methods applied to test the communication software on board the ITASAT university satellite: a case study. J. Aerosp. Technol. Manag. **6**, 447–461 (2014)
10. de Cleva Farto, G., Endo, A.T.: Evaluating the model-based testing approach in the context of mobile applications. Electron. Notes Theor. Comput. Sci. **314**, 3–21 (2015)
11. Villalobos-Arias, L., Quesada-López, C., Martinez, A., Jenkins, M.: A tertiary study on model-based testing areas, tools and challenges: preliminary results. In: 21st Ibero-American Conference on Software Engineering (2018)
12. Hashemi Aghdam, A.: Generating test adapters for ModelJunit (2017)
13. Mariani, L., Pezze, M., Zuddas, D.: Augusto: exploiting popular functionalities for the generation of semantic GUI tests with oracles (2018)
14. Legeard, B., Peureux, F., Utting, M.: Automated boundary testing from z and b. In: International Symposium of Formal Methods Europe, pp. 21–40. Springer (2002)
15. Hessel, A., Pettersson, P.: Cover-a real-time test case generation tool. In: 19th IFIP International Conference on Testing of Communicating Systems and 7th International Workshop on Formal Approaches to Testing of Software (2007)
16. Berghofer, S., Nipkow, T.: Random testing in Isabelle/HOL. In: SEFM (2004)
17. Belinfante, A.: JTorX: a tool for on-line model-driven test derivation and execution. In: International Conference on Tools and Algorithms for the Construction and Analysis of Systems, pp. 266–270. Springer (2010)
18. Veanes, M., Campbell, C., Grieskamp, W., Schulte, W., Tillmann, N., Nachmanson, L.: Model-based testing of object-oriented reactive systems with spec explorer. In: Formal Methods and Testing, pp. 39–76 (2008)
19. Barnett, M., Grieskamp, W., Nachmanson, L., Schulte, W., Tillmann, N., Veanes, M.: Towards a tool environment for model-based testing with ASML. In: International Workshop on Formal Approaches to Software Testing. Springer (2003)
20. Jacky, J.: Pymodel: model-based testing in python. In: Proceedings of the Python for Scientific Computing Conference (2011)
21. Blom, J., Jonsson, B., Nyström, S.O.: Industrial evaluation of test suite generation strategies for model-based testing. In: 2016 IEEE Ninth International Conference on Software Testing, Verification and Validation (ICSTW), pp. 209–218. IEEE (2016)
22. Amalfitano, D., Amatucci, N., Memon, A.M., Tramontana, P., Fasolino, A.R.: A general framework for comparing automatic testing techniques of android mobile apps. J. Syst. Softw. **125**, 322–343 (2017)
23. Duarte, L.M., Kramer, J., Uchitel, S.: Using contexts to extract models from code. Softw. Syst. Model. **16**, 523–557 (2017)
24. Duarte, L.M.: Behaviour model extraction using context information. PhD thesis, Department of Computing, Imperial College London (2007)
25. Cordy, J., Carmichael, I., Promislow, E.: The TXL programming language (2017). https://www.txl.ca/. Accessed 15 June 2018
26. Magee, J., Krammer, J., Chatley, R., Uchitel, S., Foster, H.: LTSA—labelled transition system analyser (2006). https://www.doc.ic.ac.uk/ltsa/. Accessed 19 June 2018
27. Blmaster: Java source: Calculator (2008). http://forum.codecall.net/topic/42522-java-source-code-calculator-app/. Accessed 15 June 2018

# Identification of UIDPs for Developing Medical Apps

Viviana Yarel Rosales-Morales[1], Laura Nely Sánchez-Morales[1],
Giner Alor-Hernández[1](✉), José Luis Sánchez-Cervantes[1],
Nicandro Cruz-Ramírez[2], and Jorge Luis García-Alcaraz[3]

[1] Tecnológico Nacional de México/I.T. Orizaba, Av. Oriente 9 No. 852 Col.
Emiliano Zapata, 94320 Orizaba, Veracruz, Mexico
viviana_rosales@outlook.com,
lauransanchezmorales@gmail.com,
galor@itorizaba.edu.mx, jsanchezc@ito-depi.edu.mx
[2] Centro de Investigación en Inteligencia Artificial, Universidad Veracruzana,
Sebastián Camacho No. 5 Col. Centro, 91000 Xalapa, Veracruz, Mexico
nicandro.cruz@gmail.com
[3] Departamento de Ingeniería Industrial, Universidad Autónoma de Ciudad
Juárez, Av. Plutarco Elías Calles 1210. Col. Fovisste Chamizal,
31310 Ciudad Juárez, Chihuahua, Mexico
jorge.garcia@uacj.mx

**Abstract.** User Interface Design Patterns (UIDPs) improve the interaction between the users and the applications through interfaces with a suitable navigability, intuitive and without restrictions in the size of the screen, to show the content. Nowadays UIDPs are very used in the development of new apps. For medical domain there are 2 types of apps: (1) apps focused on health professionals: control of appointments and files, medical encyclopedias, among others; and (2) apps focused on citizens in general: self-care of health, medical calculators, health service locators, among others. UIDPs are not considered and analyzed in the design and development process of medical apps. In this work, we carried out an analysis of UIDPS suitable for the development of medical apps, in order to identify the most used user graphical interfaces for this kind of applications. The aim of this work is to provide a guide that allows to developers incentive the use of best practices for future developments of medical apps. In conclusion, the use of best practices for the medical apps development will improve the interaction of users with the apps and it will increase the use of these apps since these are intended to make day-to-day easier of patients and their families.

**Keywords:** Development of mobile applications · Medical apps
M-Health · User Interface Design Patterns (UIDPs)

© Springer Nature Switzerland AG 2019
J. Mejia et al. (Eds.): CIMPS 2018, AISC 865, pp. 175–185, 2019.
https://doi.org/10.1007/978-3-030-01171-0_16

# 1    Introduction

In recent years, within the medical domain the use of apps has taken great importance by facilitating the day to day. For example, there are apps designed for self-care for health, which they are, apps capable of remembering a medical appointment, apps for remembering taking a medication or measuring blood glucose levels at a certain time, and that therefore could even save the lives of some people. There are apps to improve eating habits and care for patients' health, other kind of medical apps are apps to improve the doctor-patient interaction or the teaching-learning process of medicine. The medical apps are very varied, for these reasons have been classified according to the type of use given by the final users. Medical apps are classifying in two types, (1) apps for professional purposes, i.e. apps for health professionals and students of medicine or nursing and; (2) apps of general purposes for the citizen, how are the patients and relatives of the patients [1]. In another hand, software engineering is used to establish mechanisms for the software development through models, methodologies or patterns [2]. UIDPs are recurring solutions that solve common design problems. The UIDPs are standard benchmarks for the experienced UI designer [3] i.e., UIDPS are solutions to build man-machine interfaces through a graphical interface. Due to the above, different fields of the software engineering have used UIDPs to solve common design problems and one of them is software development process. The research questions that are intended to be solved in this research are aimed at: (1) how to improve the interaction of users with the apps? (2) how to facilitate the development of medical apps. From this perspective, the use of UIDPs improves the human-computer interaction by using intuitive interfaces with an adequate navigability and without restrictions on the size of the screen to adequately show the content. However, an analysis in depth to identify the most appropriate UIDPs for the development of medical apps is needed. For this reason, this paper presents an analysis of UIDPs to develop medical apps with the aim of providing a guide that helps to developer incentive the use of best practices during the design phase. A set of well-known medical apps were reviewed to identify the most used UIDPs.

This paper is structured as follows: Sect. 2 presents the state of the art about medical apps and UIDPs. In Sect. 3, the classification of mobile medical apps is presented and Sect. 4 presents the set of UIDPs identified in mobile medical apps. Finally, Sect. 5 presents the conclusions and future work is emphasized.

# 2    State of the Art

The medical apps are very broad and diverse, and the use of mobile apps in this area has increased considerably in recent years. The state-of-the-art was divided into two sections, (1) medical apps in mobile devices and (2) UIDPs in mobile apps.

## 2.1    Medical Apps in Mobile Devices

Medical apps have revolutionized the way to teach doctors and treat patients, this section presents a set of related works highlighting the use of medical apps through mobile devices in some areas of medicine. Inzunza et al. [4] described a novel proposal for an educational software platform to improve the teaching of anatomy in medical education. The research work involved the Universities of Antofagasta, Playa Ancha, Austral and Católica de Chile. The departments of anatomy that participated in the research used the educational software platform to access 2D and 3D anatomical images, videos and online multimodal practical theoretical evaluations, being able to perform usability tests with their students. The project was intended to contribute to the teaching of anatomy in different departments of anatomy. Porto Solano et al. [5] presented a framework-type model through which the information systems allowed to improve the points of care of health service providers, drastically reducing prescription errors. NFC tags were used for the identification of patients, and mobile devices for the assignment and verification of medical prescriptions. The framework provides readable prescriptions, unique access for each patient to their medical prescription, prolongation of a medical prescription by the doctor. In addition, the framework improved attention in the pharmacies of health centers. Portz et al. [6] explained that Heart failure (HF) is common in older adults and with increases in technology use among older adults, mobile applications may provide a solution for older adults to self-manage symptoms of HF. For this reason, they developed a HF symptom-tracking mobile application (HF app). The HF app is an acceptable tool for older patients with HF to self-manage their symptoms, identify patterns, and changes in symptoms, and ultimately prevent HF readmission. Gallardo López and Monroy Rodríguez [7] presented a classification of apps that help in self-care to prevent and control overweight, obesity and diabetes, since they mention that there is a growing amount of mobile applications, better known as Apps. The research was focused on different aspects of self-care and health. Ramírez López, Sánchez Mahecha and Cifuentes Sanabria [8] developed an experimental validation model of the Activ and Smca App. Activ and Smca were developed by TIGUM and used to promote the use of technology in health care. The model has a series of descriptive processes where different aspects of validation that must be met by the Apps used for health care were evaluated. In the evaluation process, the data obtained from the applications with the data obtained from certified equipment considered standard were compared. Linares-del Rey, Vela-Desojo and Cano-de la Cuerda [9] conducted a systematic review on the use of mobile applications (apps) in PD (Parkinson's disease). They found 125 applications, of which 56 were classified with potential utility in the EP, and 69 with a specific design for the EP, being 23 apps of information on EP, 29 valuation apps, 13 treatment apps and 4 valuation apps and treatment. They determined that there is a large number of mobile applications with potential utility and specific design in the EP. Nevertheless, the scientific evidence about them is scarce and of low quality. Maldonado López, De la Torre Díez, López-Coronado and Pastor Jimeno [10] presented the development and subsequent evaluation of a mobile app on Android™. The app helps in the diagnosis of eye diseases of the segment anterior of the eye, in addition to offering medical students educational content about pathologies.

## 2.2    UIDPs in Mobile Apps

On the other hand, User Interface Design Patterns have been used to solve design problems in multiple domains, due this facilitates the development of mobile applications. In this sense, Cortes-Camarillo et al. [11] developed a comparative analysis of UIDPs to generate contexts of use oriented to the educational environment. They designed contexts of use to establish which UI design patterns are recommended for a specific device by facilitating the development of multi-device educational applications. Videla-Rodríguez, Sanjuán Pérez, Martínez Costa and Seoane [12] presented a research work to determine the elements, components and key factors for the design of interactive user interfaces for augmented reality environments. Particularly, they focused in the development of virtual environments for educational applications. Quispe Rodríguez [13] carried out an assessment about users with problems in the recognition of colors, known as users color blin. From this perspective, he proposed to use interfaces with custom structures according to the needs of the users mentioned, without harming usability of the website. Espinoza-Galicia, Martínez-Endonio, Escalante-Cantu and Martínez-Rangel [14] developed a Web-based platform, mobile applications and a desktop application which allow managing information from various gadgets that monitor vital signs. Apps share characteristics among all of them, so that the reuse of source code and the saving of design time, coding and implementation are important. Apps were developed by using Microsoft .Net technology, besides using Design patterns and concepts of SOLID, CLEAN CODE and KISS through a layered development. Sánchez-Morales et al. [15] presented a software component to generate user interfaces for mobile applications using pattern recognition techniques, image processing and neural networks. The process of generating user interfaces consists of three phases: (1) image analysis, (2) configuration and (3) generation of source code. The component identifies the elements in a freehand generated image to transform each element to its equivalent source code. Cortes-Camarillo et al. [16] presented Atila, a generator of educational applications based on UIDPs. The applications developed are compatible with four operating systems: Android™, Firefox® OS, iOS™, Windows® Phone and Web. The ability to quickly generate applications is their main advantage. Markkula and Mazhelis [17] presented a generic architectural model approach to organize patterns. In the approach proposed, the identification of relevant patterns is considered as the process of reducing the set of candidate patterns by implicit restrictions in the mobile domain. These restrictions can be incorporated into a domain-specific generic architectural model that reflects the common characteristics in the particular domain solutions.

Based on the previous revision, it is possible to mention that none of the previous works performed an analysis of the UIDPs in the medical domain for the development of mobile applications. Then, Sect. 3 presents an analysis of UIDPs for mobile devices grouped by author and a classification of mobile medical apps.

## 3   Classification of Mobile Medical Apps

Currently the most popular operating systems for mobile devices are Android™, Firefox® OS, iOS™ and Windows® Phone. Each operating system has its own identity that is reflected in the appearance and behavior of each graphic element. However, all share fundamental points of view that are manifested in the design of their interfaces such as navigation, dialog boxes, notifications, among others. Different categories of UIDPs proposed by different authors were identified. For mobile devices, there are categories proposed by Neil [18], Tidwell [19], Sheibley [20], Toxboe [3] and UNITiD [21]. In the case of Smart TVs, there are categories proposed by LG Developer [22], Android™ TV [23] and Apple TV [24].

To develop an app, firstly it is necessary to determine the device and type of application to facilitate the human-computer interaction process. For medical domain, types and utilities of mobile applications are very varied. According to the last report of the IMS Health Institute [1], 65% of the available applications are focused to areas of well-being, diet and exercise. While the rest are focused on specific pathologies and treatment management, especially in the management of chronic diseases. Depending the context of use and kind of user, there are two types of medical apps: (1) apps for health professionals and (2) apps for the citizens [1]. Next, Table 1 shows the proposed classification and the medical apps of each category.

**Table 1.** Classification of medical apps

| Category | Type of application |
|---|---|
| **Professional uses:**<br>**Healthcare professionals**<br>**Medical or nursing students** | Diagnosis and treatment<br>Medical Encyclopedia<br>Control of appointments and medical records<br>Management of laboratory results<br>Medication reference<br>Medical calculators<br>Clinical communication<br>Hospital information systems<br>Medical training<br>Pharmacy<br>Evaluations<br>Tests |
| **Citizen uses:**<br>**Patients**<br>**Relatives of patients** | Location of medical entities and health professionals<br>Medical calculator<br>Management of chronic diseases<br>Guidelines and medication reminder<br>First aid guide<br>Pregnancy wheels |

A set of medical apps were analyzed to identify the UIDPs involved in the user graphic interfaces of each app. The medical apps analyzed are presented below.

## Professional Uses

*HealthTap.* This app offers medical attention "24/7" (24 h a day) from the app and the web. It has more than 70,000 associated physicians, who advise how to proceed according to the symptoms (with analytical tests, visits to the specialist, to mention but a few.). It is free, but it offers the possibility that the consultations are by video chat with the doctor for a monthly subscription [25, 26].

*FotoSkin.* It is an app championed by research staff at the University of Alcalá de Henares. The app serves as a guide for patients and doctors to diagnose skin cancer. Through a tutorial, the app teaches to take the images to generate a record and indicate the skin phototype of the user [25, 27].

*Your Medical Encyclopaedia.* The app contains detailed comprehensive illustrated information on 400+ medical conditions, which would fill more than 1000 pages of a standard medical text. All information in this encyclopedia is updated and it has been cross-referenced with a number of highly reputable sources [28].

*HealthCheck: Medical Wikipedia.* This app is an offline medical Encyclopedia that offers a rich database of medical symptoms along with a powerful search engine. The app provides information about related medical diseases, with a clean and neat design, and a user-friendly interface. Besides, the app has the option to add certain diseases to bookmarks for quicker access in the future [29].

*Patients Clinical Appointments.* This app allows registering information of the patients for generating the clinical record and for registering the appointments. The users can control his appointments dates from a mobile phone or tablet [30].

*Patient Medical Records & Appointments for Doctors.* Whit this app, doctors can manage all patient records such as personal information, medical reports, medication, visit history, clinical notes, patient history and other notes [31].

*PathoGold Laboratory Software.* This app can be used for online booking of laboratory tests appointments. The patients can search path lab near to them, make an appointment with them and get a report on their mobile phone. The doctors can see patients' lab report online [32].

*OLR - Online Lab Report.* With this app, the results of lab tests are available on the App or Web. The users can see test report given by Lab. Also, they can view and keep old reports of same lab under History. The users can download their lab reports in pdf format and record it [33].

## Uses for the Citizen

*Hospital Finder.* This app helps to locate any nearby hospital to help in emergency [34].

*Find a Doctor.* This app is a Nearest Doctor finder by using mobile phones. It uses categorization of doctors based on their specialties and facilitates appointment service through mobile, besides provides detailed information of doctors [35].

*Medical Calculators.* Pediatric Oncall medical calculators, are easy to use, available offline and immediate results for all calculation. This is a free application for all medical personnel and for patient use [36].

*MDCalc Medical Calculator.* MDCalc clinical decision tools support 35+ specialties by including: cardiology, critical care/ICU, emergency medicine, endocrinology, gastroenterology, hematology, hepatology, infectious disease, internal medicine, nephrology, neurology, obstetrics, oncology, orthopedics, pediatrics, primary care, psychiatry, pulmonology, surgery, urology, and more [37].

Next, Sect. 4 presents the result of the analysis of UIDPs in medical apps reviewed.

## 4 UIDPs Identified in Medical Apps

This section presents the result of the analysis of UIDPs in medical apps. For each medical app analyzed, all its interfaces were reviewed to identify the UIDPs used in its development. Table 2 shows the applications focused on health professionals where for each medical app the identified patterns are presented. In Fig. 1, some examples of graphical user interfaces of medical apps are presented.

**Table 2.** UIDPs identified in medical apps focused on health professionals

| Application type | Application | UIDPs |
|---|---|---|
| Diagnosis and treatment | **HealthTap** | Feedback, Datalist, Gallery, Video, Menu, Navigation Tabs, Chat |
| | **FotoSkin** | Vertical Dropdown Menu, Gallery, Map, Datalist |
| Medical Encyclopedia | **Your Medical Encyclopaedia** | Datalist, Gallery, Panel, Search, Menu |
| | **HealthCheck: Medical Wikipedia** | Dashboard, Search, Panel, Gallery, Datalist, Form |
| Control of appointments and medical records | **Patients Clinical Appointments** | Panel, Menu, Search, Datalist, Form, Pagination |
| | **Patient Medical Records & Appointments for Doctors** | Panel, Menu, Search, Datalist, Form, Pagination, Navigation Tabs, Detail View |
| Management of laboratory results | **PathoGold Laboratory Software** | Splashscreen, Feedback, Form, Panel, Menu, Dashboard, Stats, Datalist |
| | **OLR - Online Lab Report** | Login, Form, Datalist, Menu, Panel, Gallery |

Table 3 shows the result of the analysis for medical apps focused on the citizens, as well as the UIDPs identified in its graphical user interface. In Fig. 2, some examples of medical apps graphical user interfaces analyzed are presented. It is important to mention that the medical calculators can be used by patients and health professionals, it depend of the options and functionalities provided by each app.

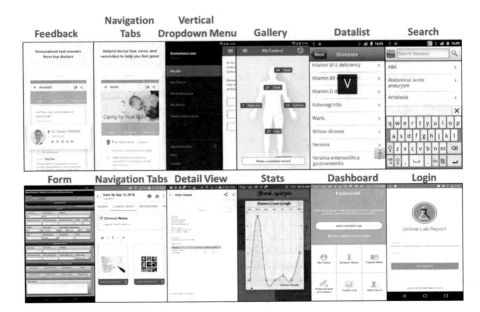

**Fig. 1.** Examples of graphical user interfaces in medical apps focused on health professionals

**Table 3.** UIDPs identified in medical apps focused on citizens

| Application type | Application | UIDPs |
|---|---|---|
| Location of medical entities and health professionals | **Hospital Finder** | Feedback, Maps, Datalist, Menu, Navigation Tabs |
| | **Find a Doctor** | Splashscreen, Menu, Vertical Dropdown Menu, Datalist, Maps, Gallery |
| Medical calculator | **Medical Calculators** | Splashscreen, Login, Form, Datalist, Menu |
| | **MDCalc Medical Calculator** | Datalist, Form, Dashboard, Panel, Menu |

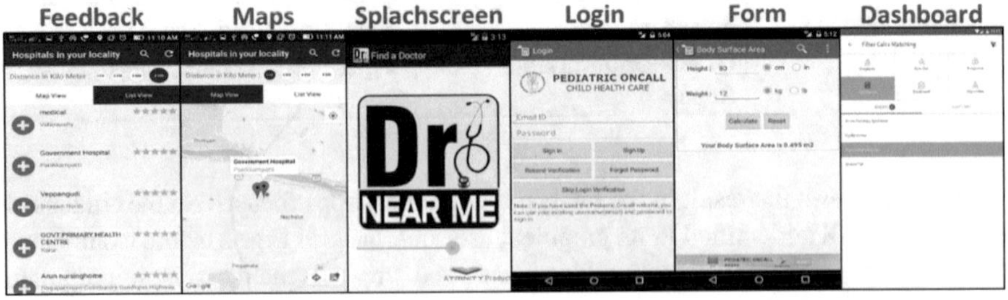

**Fig. 2.** Examples of graphical user interfaces of medical apps analyzed focused on citizens

As can be seen, medical apps are varied and they still do not have standards that define the design and development of these kind of apps. From this perspective, this work presents a significant contribution by providing an analysis by identifying the most UIDPs used in medical domain. This analysis has the objective of establishing design standards and facilitating both the use of best practices and development of mobile applications for medical purposes. This analysis reveals that the most used UIDPs in medical apps are Datalist, Login, Gallery, Video, Splashscreen, Map, Form and Menu, to name a few. Each UIDP is used according to the characteristics and functionality that the application has. For example, in applications to locate medical services the most used UIDPs are the *Maps* and *Datalist*. The Maps has support to the geolocation and the Datalist allows listing the health services units. Another example is the use of Search, Gallery and Dashboard patterns in health encyclopedias apps, where Search allows searching information about a particular pathology or medication, the gallery shows sets of images, and Dashboard presents the corresponding information.

## 5   Conclusions and Future Work

Medical apps have taken on great importance today, especially in health self-care. Due this, it is important to investigate the use of these applications in order to develop appropriate graphical user interfaces for this domain. As well as the implementation of UIDPs to solve existing problems in the design of medical apps. The analysis of UIDPs presented in this paper provides a guide to facilitate the software designer choose the appropriate UIDPs to develop a medical app considering two aspects: (1) the context of use of a given app (self-care for health, doctor-patient interaction or teaching-learning process of medicine, to name but a few) and (2) the kind of user to the app is addressed (professionals or citizens). By considering these aspects, it can be concluded that, the advantage of using UIDPs is that the apps are developed without losing quality in the design, but above all without affecting the functionality.

As future work, we are considering to expand the analysis presented, by including more medical apps that allow evaluating other UIDPs in the medical domain. We will also seek to perform analyzes in other knowledge domains such as social media and electronic commerce (e-commerce). Also, we will also seek to cover more devices in order to increase the number of UIDOs analyzed to promote the development of apps in the medical domain.

**Acknowledgments.** This work was supported by Tecnológico Nacional de México (TecNM) and sponsored by the National Council of Science and Technology (CONACyT) and the Secretarial of Public Education (SEP) through PRODEP (Programa para el desarrollo professional Docente).

## References

1. Cepeda, J.M.: Tipos y utilidades de las aplicaciones móviles de salud (2017)
2. Pressman, R.: Ingeniería del software. Un enfoque práctico (2010)
3. Toxboe, A.: UI patterns-user interface design pattern library (2012)

4. Inzunza, O., Neyem, A., Sanz, M.E., Valdivia, I., Villarroel, M., Farfán, E., Matte, A., López-Juri, P.: Anatomicis Network: Una Plataforma de Software Educativa basada en la Nube para Mejorar la Enseñanza de la Anatomía en la Educación Médica. Int. J. Morphol. **35**, 1168–1177 (2017)
5. Porto Solano, R., Porto Barceló, R., Corredor Gómez, A., Cortez Barbosa, J., Echeverri Gutiérrez, C., De los Ríos Castiblanco, J., Herrera Meza, E.: Framework ágil para el control de recetas médicas que utiliza la tecnología NFC (FARM). Rev. Lasallista Investig. **14**, 207–216 (2017)
6. Portz, J.D., Vehovec, A., Dolansky, M.A., Levin, J.B., Bull, S., Boxer, R.: The development and acceptability of a mobile application for tracking symptoms of heart failure among older adults. Telemed. e-Health **24**, 161–165 (2018)
7. Gallardo López, L., Monroy Rodríguez, G.: El autocuidado y las apps, agentes de cambio en enfermedades como sobrepeso, obesidad y diabetes. Rev. Digit. Univ. **18**, 1–12 (2017)
8. Ramírez López, L.J., Sánchez Mahecha, J.S., Sánchez Mahecha, Y.P.: Modelo de validación experimental de las aplicaciones Activ y Smca usadas para el autocuidado de la salud. Cienc. y Pod. Aéreo. **12**, 192–201 (2017)
9. Linares-del Rey, M., Vela-Desojo, L., Cano-de la Cuerda, R.: Aplicaciones móviles en la enfermedad de Parkinson: una revisión sistemática. Neurología (2017)
10. Maldonado López, M.J., De la Torre Díez, I., López-Coronado, M., Pastor Jimeno, J.C.: App Móvil de ayuda a la decisión para el aprendizaje de la asignatura "Oftalmología" en el Grado de Medicina de la Universidad de Valladolid. In: Propuesta de Innovación Educativa en la Sociedad de la Información, pp. 28–38. Adaya Press (2017)
11. Cortes-Camarillo, C.A., Alor-Hernández, G., Olivares-Zepahua, B.A., Rodríguez-Mazahua, L., Peláez-Camarena, S.G.: Análisis comparativo de patrones de diseño de interfaz de usuario para el desarrollo de aplicaciones educativas. Res. Comput. Sci. **126**, 31–41 (1870)
12. Videla-Rodríguez, J.-J., Pérez, A.S., Costa, S.M., Seoane, A.: Diseño y usabilidad de interfaces para entornos educativos de realidad aumentada. Digit. Educ. Rev. **31**, 61–79 (2017)
13. Rodriguez, A.M.J.Q.: Usabilidad web para usuarios daltónicos. Puente **8**, 71–78 (2017)
14. Espinoza-Galicia, C., Martínez-Endonio, A., Escalante-Cantú, M., Martínez-Rangel, R.: Implementación de plataforma Web y aplicaciones móviles mediante buenas prácticas usando tecnología .NET. Comunicaciones **42**, 42–49 (2017)
15. Sánchez-Morales, L.N., Alor-Hernández, G., Miranda-Luna, R., Rosales-Morales, V.Y., Cortes-Camarillo, C.A.: Generation of user interfaces for mobile applications using neuronal networks. In: García-Alcaraz, J., Alor-Hernández, G., Maldonado-Macías, A., Sánchez-Ramírez, C. (eds.) New Perspectives on Applied Industrial Tools and Techniques. Management and Industrial Engineering, pp. 211–231. Springer, Cham (2018)
16. Cortes-Camarillo, C.A., Rosales-Morales, V.Y., Sanchez-Morales, L.N., Alor-Hernández, G., Rodríguez-Mazahua, L.: Atila: A UIDPs-based educational application generator for mobile devices. In: 2017 International Conference on Electronics, Communications and Computers, CONIELECOMP (2017)
17. Markkula, J., Mazhelis, O.: A generic architectural model approach for efficient utilization of patterns: application in the mobile domain. In: Application Development and Design: Concepts, Methodologies, Tools, and Applications, pp. 501–529. IGI Global (2018)
18. Neil, T.: Mobile Design Pattern Gallery: UI Patterns for Smartphone Apps. O'Reilly Media, Inc. (2014)
19. Tidwell, J.: Designing Interfaces: Patterns for Effective Interaction Design. O'Reilly Media (2010)
20. Sheibley, M.: Mobile patterns (2013)
21. UNITiD: Android patterns (2016)

22. LG ELECTRONICS: LG developer. LG (2013)
23. Android TV: Android TV patterns (2015)
24. Apple TV: Human interface Guidelines (2016)
25. Martí, A.: Apps para diagnósticos: entre comodidad y riesgo (2015)
26. HealthTap.     https://play.google.com/store/apps/details?id=com.healthtap.userhtexpress& hl=es
27. FotoSkin. https://play.google.com/store/apps/details?id=com.wakeapphealth.fotoskin&hl=es
28. Your     Medical     Encyclopaedia.     https://play.google.com/store/apps/details?id=au.com. machealth.yme
29. HealthCheck: Medical Wikipedia. https://play.google.com/store/apps/details?id=wayback. healthcheck.com
30. Patients     Clinical     Appointments.     https://play.google.com/store/apps/details?id=com. medicalaccr
31. Patient Medical Records & Appointments for Doctors. https://play.google.com/store/apps/ details?id=us.drpad.drpadapp
32. PathoGold     Laboratory     Software.     https://play.google.com/store/apps/details?id=bms. pathogold
33. OLR     -     Online     Lab     Report.     https://play.google.com/store/apps/details?id=in.pixbit. onlinelabreport
34. Hospital Finder. https://play.google.com/store/apps/details?id=com.arogyaonline.myhospital
35. Find a Doctor. https://play.google.com/store/apps/details?id=com.dr_near_me.findadoctor
36. Medical Calculators. https://play.google.com/store/apps/details?id=Pedcall.Calculator
37. MDCalc Medical Calculator. https://play.google.com/store/apps/details?id=com.mdaware. mdcalc

# Conceptual Modelling of a Mobile App for Occupational Safety Using Process and Objectives Patterns

Oscar Carlos Medina[1]([✉]), Manuel Pérez Cota[2],
Marcelo Martín Marciszack[1], Siban Mariano Martin[1], Nicolás Pérez[1],
and Diego Daniel Dean[1]

[1] Universidad Tecnológica Nacional – Facultad Regional Córdoba, Cruz Roja
Argentina y Maestro López s/n, Ciudad Universitaria, Córdoba, Argentina
oscarcmedina@gmail.com, smarianomartin@gmail.com,
nicoperez444@gmail.com, ddean.htc@gmail.com,
marciszack@frc.utn.edu.ar
[2] Universidad de Vigo, Campus Universitario,
s/n, 36310 Vigo, Pontevedra, Spain
mpcota@uvigo.es

**Abstract.** A pattern is a model that allows the reuse of a successful solution for
the same problem in different contexts. From a Software Engineering approach,
there are different pattern types; the present work proposes the use of processes
and goals patterns to describe processes at the Conceptual Modelling phase of an
information system. We develop a study case over a process supporting an
application for occupational safety monitoring in the Public Sector. This process, due to its features and digital implementation, is considered as a repeatable
Electronic Government experience. Patterns described here are part of a pattern
catalog from the experimental phase of a research that searches to define an
analysis model for the application of Conceptual Modelling Patterns for Electronic Government systems.

**Keywords:** Patterns · Processes · Electronic Government
Occupational safety · eGov

## 1 Introduction

Electronic Government is *"the application of Information and Communication Technologies to Government processes"* [1]. Also called e-Government or eGov ("e" prefix
means electronic), it includes information systems that support processes in the Public
Sector.

In the same way that some information systems design is based in a set of business
best practices, such as management and accounting systems, the existence of Electronic
Government Good Practices, which could be used as a reference in the construction of
public software, would be advisable. One of the proposed models is the "Ibero-
American Electronic Government Charter" [2], signed by Argentina, twenty Latin-
American governments, Spain and the CLAD (Latin-American Center of Management

© Springer Nature Switzerland AG 2019
J. Mejia et al. (Eds.): CIMPS 2018, AISC 865, pp. 186–195, 2019.
https://doi.org/10.1007/978-3-030-01171-0_17

for Development), which later served as a base for an agreed "Ibero-American Model of Public Software for Electronic Government" [3]. Good government Practices are, in this context, *"every initiative and experience that help to improve the effectiveness of government actions and affect positively in citizen life conditions, achieving a measurable impact in communities"* [4].

Associated to Good Government Practices are the processes that support Electronic Government applications. These successful experiences described in a simple, accurate and standardized way, will allow taking advantage of previously gathered knowledge to a new eGov system based in the same process.

Thus, we put in consideration as a study case the inspection process for occupational safety in a government autonomous entity from Córdoba province, Argentina.

## 1.1 The Processes Patterns Origin

There is a model in Software Engineering that allows the generic description of a process such as the one stated above, and it is called "pattern". The concept of pattern was initially developed by the architect Christopher Alexander, who exposes it in his books "A pattern language" [5] and "The Timeless Way of Building" [6]. In them, he defines *"each pattern describes a problem which occurs over and over again in our environment, and then describes the core of the solution to that problem"* [6].

This pattern idea is used in software design by Gamma, Helm, Johnson and Vissides, who publishes the most important work about this subject: "Design Patterns: Elements of Reusable Object-Oriented Software" [7], in accordance to object oriented programming paradigm. With the diffusion of this book, patterns where proposed for the different phases of software development, creating specific types for each one of them, such as patterns that model business processes.

Later Eriksson and Penker enrolled in business patterns with the following pattern concept: *"a pattern is a generalized solution that can be implemented and applied in a problem situation (a context), and thereby eliminate one or more of the inherent problems in order to satisfy one or more objectives"* [8]. They published a catalog of twenty-six business patterns for business modelling encapsulated in reusable patterns (resources and rules patterns, goals patterns and processes patterns) [8].

Furthermore, goal patterns allow to define a business goals understanding goals as that what businesses seek to achieve and constitute the basis to design processes, to assign adequate resources and to adjust business rules.

Other lines of work also arise with processes patterns, which are not supported in Alexander concepts but in specific collaborative standardization efforts. One of them is, for example, from Barros, that calls process pattern to the *"common structure or architecture – with similar activities and relations- of a process that happens in all organizations and that generates the product or service that external clients demand"* [16]. Barros proposes a process redesign methodology by using patterns and the use of software to implement them, and suggests an open pattern development for the use of both public and private organizations.

Processes patterns define high quality patterns, successfully proven as facilitators of business processes modelling. Using this modelling and "business architecture" as a base, these authors propose a derivation of "software architecture" that supports them within the framework of MDD methodology ("Model-Driven Development").

This methodology searches to transform part of the Business Model in the Conceptual Model of the information system.

### 1.2    Pattern Application in Conceptual Modelling

The goal of Conceptual Modelling of a system is to identify and explain significant concepts within a problem domain, identifying the features and existing associations between them. According to Sommerville [9], Conceptual Model of a system is carried out during the first activities of "Rational Unified Process" of software development: business modelling, requirements elicitation, analysis and design. A pattern applied to Conceptual Modelling is employed to reuse knowledge and experience from previous systems encapsulated in concrete analysis and design solutions, allowing the verification and validation of functional requirements.

This proposal includes the Conceptual Model of public software [1] because it would result in several benefits. According to Beck et al. [10], the use of patterns offers the following advantages:

- Help to diminish development times
- Allow a better communication
- Help to reduce design errors, showing the essential parts of design
- Are reusable and implement the use of "good practices".

It is desirable to obtain these advantages by including patterns in Conceptual Modelling and in Electronic Government processes, proposing a study case as an example in the present work. Thus, a problem was identified and a successful solution was effectively implemented in production within a Government context, and it is described by using patterns.

Finally, it should be pointed out that reuse is one of software quality dimensions. In order to prove the effectiveness of a tool that facilitates reuse, a software quality measuring model is needed. For the specific field of Electronic Government, there are particular quality measuring models expounded on works [1, 11–14] that propose frameworks for quality assessment for electronic services and government websites.

A specific quality measuring model for e-Government systems, such as MoQGEL [13] Model, would allow evaluating if the inclusion of patterns in Conceptual Modelling optimizes its quality, reflecting it in an objective and quantifiable indicator.

## 2    Materials and Methods

The monitoring process of occupational safety considered in this work supports a mobile application called "Ubicuo" [15]. Among its clients, this software has a successful implementation in a Public Entity. Business patterns according to the definition by Eriksson and Penker [8] and Barros [16] are used to describe the software in this paper. First of all, it is graphically represented using for types of patterns: Basic Structure, interaction, layer supply and action or workflow, in a unified manner according to what is explained in [17]. It is desired to achieve an adequate understanding of the business process with this tool, in order to facilitate the graphic identification with an adequate level of abstraction that allows to represent its essential components.

Problem-Goal and Goal Decomposition patterns are additionally used, as indicated by Eriksson and Penker [8]. This figure is another simplified way of representing the business process allowing to focus on the aspects that help achieving the goals, as well as to detect problems that prevent reaching those goals.

## 2.1  Selected Study Case

EPEC (Córdoba Province Energy Corporation) is a public self-governed entity, whose main activity is to supply electric energy to Córdoba province, Argentina [18].

In order to provide this service, the covered geographic territory is formed by divisions, with different workers assigned. Each division is subdivided, as well, in areas related to different branches.

In accordance to ISO 14.001:2004 [19], that require to watch over the occupational safety and health of employees within the corporation, a monitoring process for occupational safety was developed and carried out. This monitoring process of occupational safety was implemented with the "Ubicuo" mobile application [15]. It is important to point out that old practices were improved during the process development. One of such improvements was the inclusion of technologies to a process that was previously manual and supported by paper. In this sense, the publication by OIT (International Work Organization): "*Something to point out has been the use of technological solutions, not only to improve employee registration by their employer, but also to facilitate the work of inspectors on site. In a time where the use of technology is crucial, the creation of mobile application has facilitated the supervision of occupational obligations compliance*" [20].

To manage occupational safety inspection in Public Entities. In turn, monitoring has three specific goals, or sub-goals, as shown in Fig. 1:

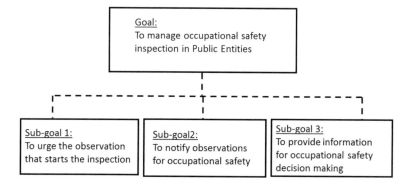

**Fig. 1.**  Goal pattern (first level) of occupational safety monitoring process.

- To urge the observation that starts the inspection
- To notify observations for occupational safety
- To provide information for occupational safety decision making.

The activities carried out during the execution of monitoring are simultaneously inter-related. In this sense, the process is dynamic, since activities are not exclusive. Instead, they are considered in relation to the whole process.

With these clarifications out of the way, we begin describing the process with the Basic Structure pattern. The graphic detailed in Fig. 2 reflects the systemic concept of the process (input, process and output) and its main goal:

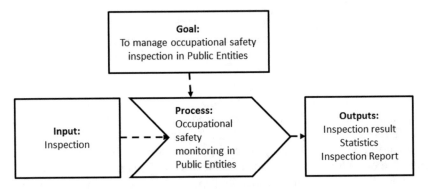

**Fig. 2.** Processes pattern (Basic Structure) of occupational safety monitoring process.

Monitoring activity consists in the execution of inspections to gather risk warnings present in work areas and in employees actions. The observations made by employees of the Public Entity, registered through this process, allow generating reports that notify and make clearly visible the problems found. The main goal of the process is achieved through these results: the management of occupational safety inspections in the organization.

## 2.2   Process Description Using Goal Pattern

At this stage it is convenient to make a pause in the construction of the processes pattern and to resume the goal pattern. The first level of the goal pattern is the graphic of Fig. 1. Sub-goals will now be further detailed by building three second level graphics.

First, it is worth noting that the process starts with the inspection activity. In this context, an inspection is the action, planned or unplanned, of an observer registering an observation, be it a "risk" event, or any observable data. Note that the wide definition broadens the basic meaning of inspection, including both negative observations (risks) and the identification of employees' good practices. This broad scope of inspection would also allow to generate not only a risks knowledge base, but also a learnt lessons one. In addition, the concept of risk is the one adopted by the OIT, referred to:

*"a physical situation with potential to cause personal harm, property or environmental damage or a combination of these"* [21].

The organization has formalized the roles of "inspector" and "monitor" to carry out the inspection activity. The former are responsible for regularly visit the facilities to register every possible observation. In order to do that, they are previously trained in the use of "Ubicuo", the mentioned informatic tool. It consists in a mobile application that allows to systematize the observations in each inspection. This system works by capturing a georeferenced photography, adding the intervention level and a categorization along a description and an action proposal for its resolution. Observed data is sent to servers through internet so that information is immediately available for consulting and processing. In a similar way, the monitor uses the tool to load observations, but in a spontaneous manner, or to confirm solutions to already registered problems. These functionalities are a good practice that responds to a determined context providing every documentary element necessary and evidentiary to solve the central problem of controlling occupational safety and health.

Several obstacles are overcome at this stage of the process, such as:

- No more out of date information is observed, from several input channels, since every observation associated with the inspection activity is managed by one unique integrated system.
- File saturation due to limited digital information, since all the information previously existing on tangible assets is now digital, for the whole process.
- Delays in result notification given that, with internet connection, all data is automatically sent to the cloud and is available for consult or processing without time or any other limitation. Work office is also reduced, related to the transcription of data between formats.
- Shortage of opportune and trustworthy information, since quality and quantity of observed data is assured by validations.

These are the problems and the solutions used by the process to fulfill sub-goal 1: Urge the observation that starts the inspection. Which is why they are added in a second level of detail of the goal pattern. Each one of them is associated to a measurable feature that allows to check if the new digitalized process improves the previous traditional one.

The process continues with the tasks carried out by another role, the moderator. From his position, he consults the observations registered by inspectors and monitors in real-time through a web page. Moderator can generate a report for each division, by selecting the observations according to different criteria: field, intervention level, categories, and state, among others. The output is generated in a PDF document. The information considered relevant to the case can also be shared with the obliged actors by a public access link. This phase of the process overcomes the delay in the generation of reports based in the volume of data to include and process, and also addresses the limitation to access these reports. Thus, the sub-goal 2 is achieved: to notify occupational safety observations. These problems and solutions are also added to the goal pattern at the second level of detail.

All the information gathered is stored in a database. This functionality allows to draft adequate criteria in order to systematize the information, and to automatically generate indicators. This way, information management enables the generation of statistics that reduce the uncertainty in decision taking. This stage of the process has the defiance to work with large volumes of historic information and a limited storing capacity. Sub-goal 3 needs to be accomplished to solve this problem: Provide information for occupational safety decision making. Thus it can be determined, for example, what kind of risks are more frequently observed or in which fields they are present.

With the third sub-goal detailed, the goal pattern for the occupational safety monitoring process is concluded.

### 2.3    Process Description Using Processes Patterns

The following patterns will be used to complete the process characterization: Basic Structure, Interaction, Layer Supply and Action Flow in a unified manner.

The top graphic section is formed by the Basic Structure pattern, previously presented in Fig. 2. The lower section shows the resources necessary to carry out the process and how they are provided, what can be seen in Fig. 3.

This graphic is read from left to right. Each assembly line is a resource or a resource set that follow the same supply flow. Each assembly line receives resources from a supplier process, represented with a line that reaches a black circle. In turn, it sends the resources to a destination process, which may be the main process or other intermediary, and is represented as a line that starts in a white circle.

The following is needed to carry out the occupational safety monitoring process in the manner proposed in this work:

- Hardware and software: provided by the Technology Management process, after being received from the Purchase Management process.
- Data Networking: supplied by the Technology Management process.
- Supplies and consumables: provided by the Purchase Management process.
- Work office and equipment: supplied by the Estate Management process.
- Human Resources and Training: provided by the Human Resources Management process.

The unified processes Pattern for the occupational safety monitoring process is finished this way.

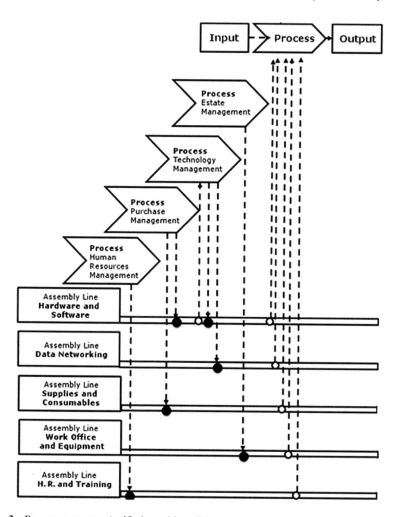

**Fig. 3.** Processes pattern (unified graphic) of the occupational safety Monitoring process.

# 3   Conclusions

For the aforementioned, it can be inferred from the Software Engineering scope that the application of Patterns allows to describe processes in the most general level of Conceptual Modelling of information systems.

In the particular case of public software, these processes support Electronic Government applications, such as de study case covered in the present work. Therefore, it is feasible to use patterns to describe eGov processes in a standardized way that allows reuse.

It has been discovered that Business Patterns are not exclusive models for systems professionals, but that they can also be elaborated by a Civil Engineer involved in the difficulties solved by the process. This sets a base for working with new study cases where different process managing disciplines collaborate.

It was also noted that the generation of these kind of patterns promotes the synthesis due to its format, helping to describe the essential parts of the process. In addition, it eases the understanding of a pattern by representing its main component graphically.

Finally, given that the effectiveness of applying patterns has been proven in concrete results, we conclude that, considering the results obtained by these patterns, they should be included in a pattern catalog for e-Government processes [11]. They can also be used in the experimentation stage of a research whose goal is to define an analysis model for the application of Conceptual Modelling Patterns in Electronic Government systems. Taking into account this case, and other eGov processes patterns, it is feasible to define a selection method for a pattern in the catalog based on the specification of the problem and it related processes.

# References

1. Medina, O.C., Marciszack, M.M., Groppo, M.A.: Aproximación descriptiva a las Buenas Prácticas de Gobierno Electrónico y a su incorporación en el Modelado Conceptual de Sitios Web Públicos de Argentina. Revista "Tecnología y Ciencia" en línea, Año 13, Artículo 100, Universidad Tecnológica Nacional, Argentina (2016)
2. CLAD Centro Latinoamericano de Administración para el Desarrollo: Carta Iberoamericana de Gobierno Electrónico. IX Conferencia Iberoamericana de Ministros de Administración Pública y Reforma del Estado, Pucón, Chile (2007)
3. CLAD Centro Latinoamericano de Administración para el Desarrollo: Modelo Iberoamericano de Software Público para el Gobierno Electrónico. Documento para la consideración de la XII Conferencia Iberoamericana de Ministros de Administración Pública y Reforma del Estado, Preparado por Corinto Meffe, Fausto Alvim y Johanan Pacheco, por encargo del CLAD - Buenos Aires, Argentina (2010)
4. Varela Rey, A.: Beneficios del intercambio de buenas prácticas municipales. Laboratorio Tecnológico del Uruguay, INNOTEC Gestión 7, 55–59 (2016)
5. Alexander, C.: A Pattern Language. Oxford University Press, New York (1977)
6. Alexander, C.: The Timeless Way of Building. Oxford University Press, New York (1979)
7. Gamma, E., Helm, R., Johnson, R., Vissides, J.: Design Patterns. Elements of Reusable Object-Oriented Software. Addison-Wesley, Boston (1994)
8. Eriksson, H.-E., Penker, M.: Business Modeling with UML: Business Patterns at Work. OMG Press, New York (2000)
9. Sommerville, I.: Ingeniería de Software 9a Edición en español. Pearson (2011)
10. Beck, D., Coplien, J., Crocker, R., Dominick, L., Meszaros, G., Paulisch, F.: Industrial experience with design patterns. In: International Conference on Software Engineering, ICSE 2018, Págs. 103–113. Technical University of Berlin, Germany (1996)
11. Medina, O.C., Marciszack, M.M., Groppo, M.A.: Proposal for the patterns definition based on good practices for the electronic government systems development. publicado en actas de CISTI 2018, AISTI y UEX Universidad de Extremadura, España (2018)
12. Rodríguez, R.A.: Marco de Medición de calidad para gobierno electrónico. Aplicable a sitios web de gobiernos locales. Publicado en actas de XIV Workshop de Investigadores en Ciencias de la Computación - WICC 2012, Posadas - Misiones, Argentina (2012)

13. Sá, F.A., Rocha, A.: Qualidade do Governo Eletrónico. Modelo MoQGEL. Ed. Sílabo (2017)
14. Sá, F.A., Rocha, A., Pérez Cota, M.: From the quality of traditional services to the quality of local e-Government online services: a literature review. Government information Quarterly N° 33 (2016), 149–160, Elsevier Inc. (2015)
15. Ubicuo: Nuestros productos (2018). http://www.ubicuo.com.ar. Extraído el 05/09/2018
16. Barros, O.: Rediseño de proceso de negocios mediante el uso de patrones. Mejores prácticas de gestión para aumentar la competitividad. Dolmen Ediciones S.A., Chile (2000)
17. Marciszack, M.M., Castro, C., Sánchez, C., Delgado, A., Garnero, A.B., Horenstein, N., Fernández, E.: Una experiencia en la aplicación de Patrones de Negocio. publicado en actas de CONAIISI 2016, Red RIISIC, CONFEDI y UCASAL Universidad Católica de Salta (2016)
18. EPEC: ¿Quiénes somos? (2018). https://www.epec.com.ar/institucional.html. Extraído el 05/09/2018
19. ISO: ISO 14000 family - Environmental management (2018). https://www.iso.org/iso-14001-environmental-management.html. Extraído el 05/09/2018
20. Organización Internacional del Trabajo: Notas sobre tendencias de la inspección del trabajo, FORLAC, OIT (2015)
21. Organización Internacional del Trabajo: Inspección de seguridad y salud en el trabajo – Modulo de formación para inspectores. OIT (2017)

# PulAm: An App for Monitoring Crops

Alejandra Perez-Mena[1], José Alberto Fernández-Zepeda[1],
Juan Pablo Rivera-Caicedo[2], and Himer Avila-George[3(✉)]

[1] Centro de Investigación Científica y de Educación Superior de Ensenada,
Ensenada, B.C., Mexico
perezmena@cicese.edu.mx, fernan@cicese.mx
[2] Cátedras CONACyT, Universidad Autónoma de Nayarit,
Tepic, Nayarit, Mexico
jprivera@uan.edu.mx
[3] Centro Universitario de los Valles, Universidad de Guadalajara, Ameca,
Jalisco, Mexico
himer.avila@academicos.udg.mx

**Abstract.** In this paper, we introduce PulAm a mobile application based on the
Android operating system, which was designed as a support tool for the monitoring process of different crops. As a case study, we introduced the monitoring
of sorghum crops, specifically against the "*yellow sugarcane aphid.*"

**Keywords:** Mobile application · Sorghum · Yellow sugarcane aphid

## 1 Introduction

Agriculture is a primary activity of great relevance to the human being from the
economic, social and environmental point of view since it generates a large number of
jobs and food. It also helps to conserve soil and biodiversity. Sorghum is the second
most important crop in the Mexican economy. It mainly grows in the Mexican states of
Tamaulipas, Guanajuato, Sinaloa, and Nayarit [1]. The pest of the *yellow sugarcane
aphid* appeared in the north of Tamaulipas in 2013 and severely damaged sorghum
crops. Since then, sorghum production has decreased between 30% and 100% in
several areas [2].

The yellow sugarcane aphid, *Melanaphis sacchari*, is an insect classified in the
*Aphididae* family. This insect has gregarious habits, i.e., lives in colonies, specifically
on the underside of the leaves of the invaded plants, see Fig. 1. Initially, the colonies
appear at the base of the sorghum plant and gradually move to the top, which can even
infest the panicle of sorghum in flowering [2]. A sorghum plant with yellow sugarcane
aphid pest shows changes in its leaf color because the aphid feeds on the sap of the
plant. This process causes loss of nutrients and stress in the plant. The insect also
secretes a sugary substance on the surface of the leaves which causes the appearance of
sooty mold and promotes the spread of viruses such as the sugarcane mosaic or the
sugarcane yellow leaf. However, the most critical damage in the crop is the reduction of
the quality of the sorghum seed and the loss of harvest yield [3].

© Springer Nature Switzerland AG 2019
J. Mejia et al. (Eds.): CIMPS 2018, AISC 865, pp. 196–205, 2019.
https://doi.org/10.1007/978-3-030-01171-0_18

**Fig. 1.** Sorghum leaves infested by yellow sugarcane aphid.

Given the economic importance of sorghum crops in Mexico and the great damages caused by the yellow sugarcane aphid, in 2015, the National Service of Health, Safety, and Agri-Food Quality (SENASICA), through the State Committees for Vegetal Health, authorized the operation of the Campaign against this pest in sorghum. This campaign includes actions such as monitoring, application of biological, chemical, and cultural control, with an emphasis on training for producers.

In the state of Nayarit, its State Committee for Vegetal Health (CESAVENAY) is responsible for implementing this campaign. The activities carried out by the CESA-VENAY technicians during these inspections are the following. First, evaluating if the pest is present in the crop, if this is the case, determining its degree of infestation. Second, collecting information related to the plantation and the crop (date of sowing, phenological stage, level of damage, and geographic information). Third, training sorghum producers in the management of the crop and the prevention and treatment of the pest. The technicians of CESAVENAY gather manually all the information related to the process of monitoring sorghum crops in logbooks. Then, they capture data in a web platform; this process has caused many inconsistencies in the integrity of the information, mainly with geographic coordinates, the phenological stage of the crop, and its degree of infestation.

Previously, some research efforts have developed mobile applications related to precision agriculture and support for decision making to demonstrate its usefulness and versatility in the monitoring of crop development. One of them is FarmBeats [4], an application being part of an IT solution through which the user can visualize the results of analysis of soil and health variables of a crop. This IT solution contemplates scenarios where there is no Internet connection and how it stores its data. The work in [5] describes a system that includes different tools (web services, an Android application, and a database). With the Android app, the producer can add all the information related to the system, demonstrating the usefulness of an application in the information collecting process information. The paper in [6] describes an Android app that is responsible for the acquisition of crop information through sensors and a database, along with other components of the system, it helps with the decision making related to the crop development.

In this work, we describe the design and implementation of PulAm, a mobile application to automatize the monitoring process of different crops, mainly during the capture of field information. The application runs on the Android operating system. With this application, a technician can identify crops, automatically capture their geographic information, and based on his/her experience, dictate the degree of infestation. Our case study focuses on studying the impact of the yellow sugarcane aphid in sorghum crops.

## 2   Traditional Monitoring Process

CESAVENAY has a team of technicians who are in charge of monitoring the pest of the yellow sugarcane aphid in sorghum crops. They design weekly planning of the inspections to be carried out in different sorghum plantations. They make this design according to the historical data of the inspected crops, the presence of the pest, and its degree of infestation. The technicians also divide the inspections to crops according to their geographical distribution and to the management and social activities that they must carry out.

On an inspection day, the technician arrives early at their inspection zones and distributes chemical/biological control to the producers. Then, the technician reviews the crops according to his/her planning. During this activity, they register new plantations if the producer requires it. The technician evaluates the degree of infestation in a certain number of plants. He/she records it in the crop logbook along with the following information: coordinates of the inspection point (with the GPS of his/her mobile device), name of the pest, degree of infestation, number of inspected plants, phenological stage of the crop, variety of sorghum, inspection date, among other data. Finally, the technician manually enters the obtained data during the inspections and specific management information to a CESAVENAY's computer platform.

The most frequent problems detected in this information capture process are the following:

- Technicians may cause errors when entering the inspection information to their logbook and from this to the computer platform.
- Technicians do not always record the new coordinates of the inspection point, but they associate it with the coordinates of a previous nearby point.
- There is no way to verify that technicians performed the inspections at the coordinates they registered.

## 3   Proposal of a Mobile Application

This work introduces a mobile application called `PulAm` to address the problems described previously and to automatize the monitoring process of crops. This section describes the objective of this application, its technical description, and its main components.

### 3.1   Objectives of PulAm

The main purpose of `PulAm` is to support technicians to capture all the information related to the level of infestation in crops without worrying about technical details of the process, such as geographical validations and data integrity. `PulAm` also has the following secondary objectives: (a) allowing automatic and reliable capture of inspection information, and (b) consulting and editing information regarding inspections.

### 3.2   Technical Description of PulAm

We developed `PulAm` for Android 4.0 (KitKat). The mobile device in which `PulAm` runs must have a GPS and an Internet connection. The mobile application needs to connect to a Web server to update the changed, erased or added information in its database. Figure 2 shows the architecture of `PulAm`. Besides this application, the mobile Android device also hosts an `SQLite` database from which `PulAm` obtains data. This device connects to the server, through the Internet and a `PHP` service, to update the information modified or added to a `MySQL` database that performs other activities.

### 3.3   PulAm Modules

`PulAm` consists of four modules: (1) capture of inspection; (2) database management; (3) crop verification; and (4) database synchronization. `PulAm` also allows users to customize option of map design and to facilitate access to data. Figure 3 shows the main screen of `PulAm`, from which users can access the application modules.

*Capture of Inspection*
In this module, the technician can capture the degree of infestation of the pest. It automatically captures the coordinates, date, time, and completes the record with the information of the latest added inspection. The technician can modify the latter information to record the current inspection. The technician does not have permission to edit the coordinates and the datetime of the inspection. Figure 4 shows three screenshots related to this module, as well as its visual elements.

*Database Management*
`PulAm` takes information from the `SQLite` database hosted on the mobile device. Figure 5 shows the relational diagram of this database.

The database management module is responsible for the consulting, editing, deleting, and adding new registers of inspections, pest, crops, technicians, and

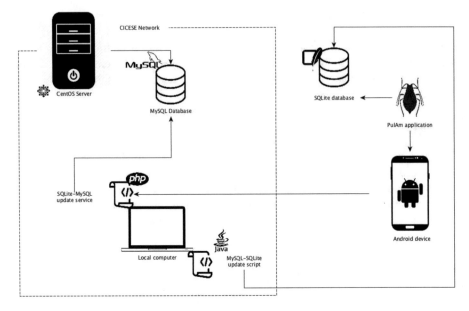

**Fig. 2.** The architecture of PulAm.

**Fig. 3.** The main screen of PulAm with its modules: (1) capture of inspection; (2) database management; (3) crop verification.

plantations. With this functionality, PulAm allows handling other pests and crops different from the sugarcane yellow aphid and sorghum, respectively. The management of each table has restrictions that are related to the priority of the information and how the modification affects other tables. To minimize connectivity problems, we implemented a local database on the mobile device instead of a web service.

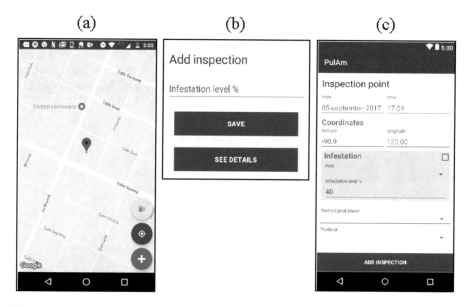

**Fig. 4.** Screens of the capture of inspection module. (a) Selection of the crop. (b) Capturing the degree of infestation. (c) Edition of the inspection and summary.

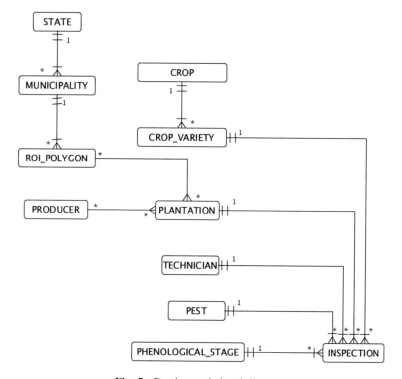

**Fig. 5.** Database relational diagram.

*Crop Verification*

In the crop verification module, the user can select who owns a plantation. It also shows who is currently working on the plantation, and what type of crop it has. We implemented this module because there are plantations whose current producer does not match the one registered in the database, see Fig. 6. Therefore, this module verifies such information is correct.

**Fig. 6.** The screen of the crop verification module.

*Database Synchronization*

This module is responsible for synchronizing all deleted, edited, and added information through `PulAm` when the mobile device connects to the Internet. Each time a user changes something in the database, the query enters into a queue. Once the mobile device connects to the Internet, this module sends this information to the server through the `HTTP` protocol which captures it with a `PHP` service. Finally, the server executes the query in the `MySQL` database. Updating the `MySQL` database to `SQLite` requires a script in Java. We did not install a `MySQL` database on the mobile device because it demands a lot of resources in comparison with an `SQLite` database [7]. This module is useful when there is no available Internet connection; the information stays in the queue until it can be synchronized.

*Personalization of the Application*

`PulAm` allows users to customize the following characteristics: type of map to be used (by default the map changes color by the daytime) and the technician who will use the application, see Fig. 7.

## 4   Field Tests

To test `PulAm`, we carried out a monitoring campaign, which consisted of visiting 40 sorghum plantations; see Fig. 8. With the information collected, were identified some improvements needed by the application.

**Fig. 7.** PulAm customization screens.

**Fig. 8.** Sorghum monitoring campaign.

Next, we performed functionality tests on the four main modules of PulAm. From these tests, we made the following improvements:

- The user had to select his/her name manually every time he/she wanted to add new inspections. Subsequently, we added an option in the preferences menu, so the user must do it only once.
- We configured the mobile device in such a way that in the absence of a GPS signal, the device can request the most accurate data available via Wi-Fi and network provider's cell towers.
- We added a dialog box in the capture of inspection module to ease the addition of a new inspection. Now, the user only must add the degree of infestation while the application takes the rest of the information from the last added inspection, information that is stored automatically in the shared preferences file. Finally, the user can edit this information according to the data of the current inspection.
- PulAm initially used a series of points stored on the database that together formed a polygon represented a plantation. The application had to draw all the points to form the polygon, using high computational resources. Now, PulAm reads polygons as a geographical layer that it draws on the map quickly.
- We added additional functions in the database management module.

## 5   Conclusions and Future Work

PulAm is a prototype through which the CESAVENAY technicians can record the results of their crop inspections in an agile way and minimizing errors. They may also can access historical information about the properties and previous inspections.

We are planning to carry out acceptance tests of PulAm with the CESAVENAY technical staff. We are also planning to design a module in which the user can take one or more photographs of the sorghum plants with his/her mobile device and the application automatically determines its degree of infestation. With this module, the user would have a more standardized measurement of the degree of infestation.

**Acknowledgments.** We thank the staff of CESAVENAY for all their help, both technical and logistical for the data collection in the field. Alejandra Perez-Mena thanks CONACyT for the financial grant provided through the scholarship # 613382.

## References

1. SAGARPA: Atlas agroalimentario México. Secretaría de Agricultura, Ganadería, Desarrollo Rural, Pesca y Alimentación (2014)
2. Rodríguez-del-Bosque, L.A., Teran, A.P.: Melanaphis sacchari (*Hemiptera: Aphididae*): a new sorghum insect pest in Mexico. Southwest. Entomol. **40**(2), 433–434 (2015)
3. SENASICA: Pulgón amarillo Melanaphis sacchari (*Zehntner*). Servicio Nacional de Sanidad, Inocuidad and Calidad Agroalimentaria (2014)

4. Vasisht, D., Kapetanovic, Z., Won, J., Jin, X., Chandra, R., Kapoor, A., Sinha, S.N., Sudarshan, M., Farm, S.C.: FarmBeats: an IoT platform for data-driven agriculture. In: USENIX Symposium on Networked Systems Design and Implementation, vol. 14, pp. 14–20 (2017)
5. Liopa-Tsakalidi, A., Tsolis, D., Barouchas, P., Chantzi, A.-E., Koulopoulos, A., Malamos, N.: Application of mobile technologies through and integrated management system for agricultural production. Procedia Technol. **8**, 165–170 (2013)
6. Montoya, F.G., Gómez, J., Cama, A., Zapata-Sierra, A., Martínez, F., De La Cruz, J.L., Manzano-Agugliaro, F.: A monitoring system for intensive agriculture based on mesh networks and the android system. Comput. Electron. Agric. **99**, 14–20 (2013)
7. DigitalOcean: SQLite vs MySQL vs PostgreSQL: A Comparison of Relational Database Management Systems. https://www.digitalocean.com/community/tutorials/sqlite-vs-mysql-vs-postgresql-a-comparison-of-relational-database-management-systems

# A Computational Measure of Saliency of the Texture Based a Saliency Map by Color

Graciela Lara[1(✉)], Adriana Peña[1], Carlos Rolon[1],
Mirna Muñoz[2], and Elsa Estrada[1]

[1] CUCEI of the Universidad de Guadalajara,
Av. Revolución 1500, Col. Olímpica, 44430 Guadalajara, Jalisco, Mexico
{graciela.lara,elsa.estrada}@academicos.udg.mx,
adriana.pena@cucei.udg.mx, carlosagrolon@gmail.com
[2] Centro de Investigación en Matemáticas,
Avda. Universidad no. 222, 98068 Zacatecas, Mexico
mirna.munoz@cimat.mx

**Abstract.** Understanding what attracts the human eye has always interested scientists; this represents the saliency concept. In this context, texture is a visual feature of the external surfaces of objects that can influence the senses, creating a force of attraction or rejection. When the texture of an object attracts the viewers' attention it represents a salient feature. However, measuring how salient is an object texture is not trivial. In this paper we propose a computational measure of saliency of the visual texture, using a salience map by color. This measure of saliency is part of a computational model aimed to the selection of an appropriate reference objects to facilitate the location of objects within a 3D virtual environment. In order to verify this computational measurement, an experiment was conducted with 40 people, 30 male and 10 female, to understand to which extent the proposed measure of saliency matches with the people's subjective perception of saliency. Results show a good metric performance.

**Keywords:** Texture · Attention · Saliency map · Color

## 1 Introduction

When we perceive an object we see it as an entire entity where several features are configured such as: color, size, texture, orientation, position, or luminance; its visual attraction is largely determined by these basic features.

In the artistic world, texture takes special importance in the areas of design, sculpture and decoration. Texture as a feature of the surface of an object can be classified into two main categories: visual texture and tactile texture. In this article we focus our attention on the visual texture, treated from a saliency map by color.

The visual texture is a perception, whose appearance depends on the effects of the contrast (light and shadow). The visual texture has a two-dimensional focus. The texture decorates a surface and is subject to the figure or object. The texture itself is only an aggregate that can be removed without affecting the figures and their interrelationships in the design. It can be drawn by hand or obtained by special resources, and

© Springer Nature Switzerland AG 2019
J. Mejia et al. (Eds.): CIMPS 2018, AISC 865, pp. 206–215, 2019.
https://doi.org/10.1007/978-3-030-01171-0_19

it can be regular or irregular, in any case it maintains a certain degree of uniformity. The texture of an object influences the senses of the humans, creating an attraction.

The saliency is a key concept of psychology that has been applied in information technology for object analysis and vision [1], as well as to locate salient areas in images (e.g. [2–7]). Lahera et al. [8] define visual saliency as the automatic and subliminal process of bottom-up visual discrimination, whereby certain stimuli stand out from the perceptual field and attract attention. In other words, it is the higher-order mental process by means of which certain perceived or mentally represented objects attract the focus of attention, including thinking and behavior. In this sense, visual saliency refers to bottom-up processes that render certain image or object regions more relevant: for example, image regions with different features from their surroundings (e.g., a red rose among several white roses).

Attention or visual attraction is linked to the concept of saliency. In this sense, the visual saliency of an object depends on the interaction of its basic features [9–12], with respect to other objects.

The ability of the human visual system to detect the visual saliency of an object, or a set of objects, is extraordinarily fast and reliable, the computational modeling of this basic intelligent human behavior still remains as a challenge [2].

Given the importance of saliency to diverse aspects of vision science and its applications, good measures of salience are clearly desirable. So that, in recent decades has attracted the researchers' attention (e.g. [13–17]). Texture on objects is often used as cue for quickly selecting important aspects of a scene [18]. Furthermore, texture is considered an important cue for attention, segmentation and object detection; only few saliency models currently exploit texture [19].

Syeda-Mahmood [18] developed an algorithm to measure the salience of the texture, extracting the relatively bright and dark regions of the texture within an intermediate gray background as capturing in a set of four binary maps. The author analyzes their interaction in terms of perceptually highlighted attributes.

Stürzel and Spillmann [16] studied the perceptual saliency from texture, estimating the magnitude and the reaction time, from various kinds of texture contrast between target and background, to predict the time of fading and stimulus patterns. They used three different classes of texture targets with constant border lengths and location on similarly textured backgrounds. They used a fourth stimulus as a control. Targets differed from their surrounds either in orientation, shape, or order (regularity). They found that the physical strength of the texture contrast correlates with the fading time according to the perceptual prominence of the stimulus: the greater the prominence, the greater the fade time.

Kadir and Brady [20] developed a method for the characterization of textures, calling Scale Saliency Descriptors, where textures are represented through the PDF of their salient scales. The PDF is represented as a histogram. Their approach considers each pixel location, where is calculated the entropy from the PDF of a descriptor; they used a circular window to sample the image. The process is repeated for increasing scales; scales are selected at the reached entropy, then the entropy value is weighted by some measure of the self-dissimilarity in scale-space of that feature. The authors applied their method for supervised texture classification and unsupervised segmentation tasks, using a variety of synthetic and real data.

Wang et al. [21] proposed an algorithm to detect outgoing objects in the selective contrast. The selective contrast is the process to explore intrinsically the information of distinguishable components such as color, texture and location. They extracted the texture by the outputs of a set of filters. The outputs comprise a wide range of high-dimensional spaces that are grouped by k-means, so that the centers of the clusters can be treated as representative textures.

A good salience method based on texture was developed by Terzić et al. [22]. They characterized the local texture of each pixel, through a 2D spectrum obtained from Gabor filters. Next, they applied a parametric model, to describe the texture of each pixel with the combination of two Gaussian 1D approximations. From this they obtained a model composed of four parameters. The four parameters are used to then apply the Difference-of-Gaussian blob detection. This method was bio-inspired to help explaining the role of texture in the early saliency processing, and how it drives saccadic eye movements to objects.

Zhang et al. [23] proposed a model to detect salience, using color and texture as characteristics. They used an algorithm called SLIC superpixel to perform segmentation on the image. The SLIC superpixels algorithm uses the k-means clustering approach to generate superpixels in CIELAB space, considered fast and memory management efficient. Furthermore, they used a method of region contrast adaptation weights, calculating the salience of two maps: one per color and the other for texture. Finally, they introduce an effective and logical fusion method with the previous maps to obtain a better map of saliences.

Finally, Terzić et al. [19] proposed another texture salience model, in which they consider the texture of each position that is characterized, with the two-dimensional local power spectrum that is obtained from the Gabor filters using several scales and orientations.

From this brief review it follows that there is no unanimous proposal, toward the goal for which is directed our computational model of saliency of the texture. Thus, it should be noted that the salience measures proposed in these works were designed and tested to capture the attention under particular conditions. The results found show an attention capture, according to the desired objective.

Texture is one of the most important visual elements. We use textures to enrich works, represent reality, add expressiveness or sensations and to create visual relationships between the elements of an artistic work. The texture of objects from its visual aspect, without considering the tactile aspect, can be analyzed; likewise, regardless its volume saliency can be studied. Texture is an important feature for many tasks related to vision. However, it is not used in most models of relevance, because textures are a complex aggregations of shapes or colors, with variations or irregularities, whose contrast effects (light and shadow) have different intensity, see Fig. 1; this features make textures difficult to treat.

Therefore, calculating the saliency measure on textures definitely remains a challenge. The reason behind this challenge is mainly due to the lack of a simple and well-defined model for identify saliency.

Our proposal for the measurement of texture saliency may help researchers to test and improve a number of vision models based on any textures, and to better predict the

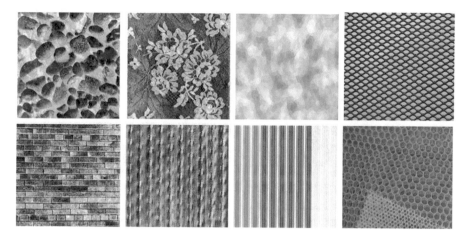

**Fig. 1.** Samples of textures

importance to human observers of different kinds of colors degradation. In the next section, our measure of saliency is described in detail.

## 2  A Measure of Saliency of the Texture

Since the visual texture is strictly two-dimensional, our measure is based on the central idea of a saliency map. The saliency map is calculated based on the saliency by color of each pixel. We applied this measure of saliency to the texture of the image, which wrap an object. To do so we considered the following steps:

(1) The color of each pixel of the texture image is obtained in the RGB coordinate system, using a technique inspired on the rendering of 3D graphics within a 2D image. The technique consists in launching several rays, one for each pixel in the texture image.

   This process is repeated as many times as pixels contains the image. It was not possible to obtain a sum of the pixels by color, since the images of the textures used were very varied, with little uniformity, and when recovering the coordinates of each color intensity difference was detected between the pixels from sumptuously the same color. For example, a bright red color could have an R value of 246, a G value of 21, and a B value of 51. The brightest red you can get an R value of: 255, G: 0, B: 0.

(2) After, we transformed the RGB coordinates to CIE Lab format. To calculate the saliency by color of each pixel, we used the process proposed in [24], which is summarized in the following three step:

   1. Choungourian [25] proposed a group of eight colors with major stimuli to the human visual system. These colors set is red, orange, yellow, yellow-green, green, blue-green, blue, and purple. Table 1 presents the list of these colors with their numerical representation in the RGB and CIE-Lab systems.

**Table 1.** List of salient colors according to Choungourian

| Color | R | G | B | CIE L* | a* | b* |
|-------|---|---|---|--------|-----|-----|
| Red | 255 | 0 | 0 | 53.23 | 80.10 | 67.22 |
| Orange | 255 | 112 | 40 | 64.16 | 51.02 | 62.62 |
| Yellow | 255 | 255 | 0 | 97.13 | -22.55 | 94.48 |
| Yellow_green | 154 | 205 | 50 | 76.53 | -37.99 | 66.58 |
| Green | 0 | 255 | 0 | 87.73 | -86.18 | 83.18 |
| Blue-green | 13 | 152 | 186 | 58.11 | -20.81 | -26.94 |
| Blue | 0 | 0 | 255 | 32.30 | 79.19 | -107.86 |
| Purple | 102 | 2 | 153 | 27.20 | 59.25 | -56.42 |

The distances between the color of each pixel of the texture with respect to the group of the proposed colors by [25], is calculated using Euclidean distances.

2. After, the salient color with less distance to the color of the pixel is selected, and a saliency value is assigned to each of its 3 coordinates (L, a, b) according to the next ranges:

   - '1.00' for distances of less than 5 units
   - '0.75' for distances bigger than 5 and less than 10 units
   - '0.50' for distances bigger than 10 and less than 15 units
   - '0.00' for distances bigger than 15 units

   This is a rough categorization for color saliency, and we are aware of it, but to our knowledge there are no graduating intermediate values on this regard.

3. Then, the three saliency values are averaged to get the saliency by color for each pixel.

(3) Finality, the value of saliency by texture image *(sti)* is calculated, by the average of the sum of saliency by pixel *(Σspi)* and the total of pixel *(tpi)* of texture image, as is applied in Eq. (1)

$$sti = \sum spi/tpi \tag{1}$$

This measure provides a rough estimation of the saliency of a texture image. Given that, our final goal is to be able to identify the most salient 3D objects within a 3D virtual environment. This type of saliency is strongly dependent on the context in which the objects are located; a normalization process is applied over the previously computed value. The object with the highest value of saliency has a value of '1' and the rest a proportional value. Normalized saliency by texture is represented by $sti_{[0-1]}$ value.

## 3   Experimental Evaluation of the Texture Saliency Metric

With the aim to verify our general hypothesis: Is it possible to achieve the salience detection process of the human visual system with our saliency approach?, the next described experiment was designed and carried out.

## 3.1    Method

**Participants.** A total of 40 participants (10 females, 30 males) undergraduate students of the Department of Computer Science of the Universidad de Guadalajara, with ages in the range of 18 to 25 years, and normal vision, voluntarily participated.

**Materials, Devices and Situation.** The experiment was carried out in a laboratory with suitable lighting condition. Our experiments are performed on a Desktop Gamer Lenovo™, model IdeaCentre Y700 34ISH, with a processor Intel ® Core™ i7, 3.6 GHz DDR4 2400 MHz, 8 GB memory, using a mouse and a keyboard. A computational application was developed to implement and test the metric using the Unity 3D™ platform, with some scripts created in C# programming language. Results were automatically recorded in a .csv (comma-separated values) file and afterwards statistically analyzed using the SPSS™ (Statistical Product and Service Solutions) application.

**Design and Procedure.** In our experiment, we provide ten scenarios (called trials), each presenting four 3D objects in the form of spheres with a texture on the surface (see Fig. 3). Forty texture images were extracted randomly from two databases that are public available in the World Wide Web [26, 27]. (1) The Multiband Texture (MBT) database, which is a collection of 154 color images. Images from the MBT database have a rich chromatic content that does not have discriminative value, yet it contributes to form textures; and (2) the Describable Textures Dataset (DTD) that is an collection of textural images in the wild, with a series of human-centric attributes, inspired by the perceptual properties of textures. The set of selected texture images for this experiment were adjusted to have the same size scale. Likewise, the set of texture images were edited with the GIMP (GNU Image Manipulation Program) cross-platform. Each texture image was edited using the color group proposed by Choungourian [25] and with other random colors, the time for the edition process of each texture image was of 20 min. Figure 2 shows the set of the texture images used in the experiment.

**Fig. 2.** View of the forty textures used

**Table 2.** Classification of textures by saliency value

| Saliency Value | Classification |
|---|---|
| Bigger than .75 and less than 1 | High |
| Bigger than .50 and less than .75 | Medium |
| Bigger than .25 and less than .50 | Low |
| Less than .25 | Null |

Likewise the set of texture images for this experiment were labeled and classified, according to their salience value, as show the Table 2.

The task of the experimental participants was to place the four objects provided on each trial on an empty platform in front of them (see Fig. 3), ordering them from left to right according to their texture saliency. This concept was explacated to them as "the capability of the object's texture to attract their attention". Therefore, the most outstanding object by its texture should be placed to the left. To make sure that the participants had understood the instructions well were given a brief demonstration of the system on how to place each object on the platform. Also, participants were told that they could make all necessary changes, before they confirmed the final order of objects for each trial. Participants were asked to provide basic personal information as their, age and gender, within the system. Each subjects lasted about 5 min to complete the ten trials. The experiment was sufficiently brief that no rest periods were required.

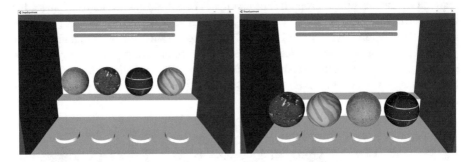

**Fig. 3.** View of an experimental system trial

### 3.2 Statistical Analysis of Results

With the objective of assessing our general hypothesis of how similar is our metric to the perception of saliency of the participants, we conducted same statistical analyses based in [28]. We analyzed the following aspects:

1. *The choice of the first or second object as the most salient by its texture.* Because the salience of the texture of an object is a very subjective characteristic, because the sensory properties come into play, and given that there is no precise order with which to compare the given order of our metric and each of the given orders of the participants, the comparison required to be flexible enough to admit variations, but

at the same time capable to give information about the performance of the metric. We decided to measure the extent to which the first or second most salient objects, according to our metric of saliency, matched the object placed by the participants in the first position. This condition helped us to identify if the most salient objects to the human visual system corresponds with high-valued objects by our metric.

2. *Comparison of the mode with respect to the metric.* In this second statistical analysis, each trial was compared with the order given by our metric. The mode of the most voted object in each position for each trial was obtained, counting the number of times that an object was placed in each of the four positions within each trial. Then, the distance between each pair of objects (the one given by the metric and the most voted one) in each position of each trial was computed, according to the value assigned to each object by our metric.

Nevertheless, although our statistical analyses are the same of [27], the purpose of the experiment is different. Our main interest in these statistical analyses is the study of saliency in textures.

## 4 Results and Discussion

The results of the first statistical analysis show that our metric, help to predict the objects with the texture that humans tend to perceive as the most salient. The number of possible orders for the 4 objects of each trial is 24, but only half of them fit with the

**Table 3.** Order of the textures in each trial: (a) based on our metric of saliency and (b) based on the mode (When the order is the same in both, the cell is highlighted in green)

(a)

| Trial | Pos_1 | Pos_2 | Pos_3 | Pos_4 |
|---|---|---|---|---|
| Trial_1 | **TextureHS-01** | TextureMS_01 | TextureLS_01 | **TextureNS_01** |
| Trial_2 | TextureHS_02 | **TextureMS_02** | TextureLS_02 | **TextureNS_02** |
| Trial_3 | TextureHS_03 | TextureMS_03 | **TextureLS_03** | **TextureNS_03** |
| Trial_4 | TextureHS_04 | TextureMS_04 | **TextureLS_04** | **TextureNS_04** |
| Trial_5 | **TextureHS_05** | **TextureMS_05** | **TextureLS_05** | **TextureNS_05** |
| Trial_6 | TextureHS_06 | TextureMS_06 | TextureLS_06 | TextureNS_06 |
| Trial_7 | TextureHS_07 | **TextureMS_07** | TextureLS_07 | **TextureNS_07** |
| Trial_8 | TextureHS_08 | **TextureMS_08** | TextureLS_08 | **TextureNS_08** |
| Trial_9 | TextureHS_09 | **TextureMS_09** | TextureLS_09 | **TextureNS_09** |
| Trial_10 | **TextureHS_10** | TextureMS_10 | **TextureLS_10** | TextureNS_10 |

(b)

| Trial | Pos_1 | Pos_2 | Pos_3 | Pos_4 |
|---|---|---|---|---|
| Trial_1 | **TextureHS_01** | TextureLS_01 | TextureMS_01 | **TextureNS_01** |
| Trial_2 | TextureLS_02 | **TextureMS_02** | TextureHS_02 | **TextureNS_02** |
| Trial_3 | TextureMS_03 | TextureHS_03 | **TextureLS_03** | **TextureNS_03** |
| Trial_4 | TextureMS_04 | TextureHS_04 | **TextureLS_04** | **TextureNS_04** |
| Trial_5 | **TextureHS_05** | **TextureMS_05** | **TextureLS_05** | **TextureNS_05** |
| Trial_6 | TextureNS_06 | TextureHS_06 | TextureMS_06 | TextureLS_06 |
| Trial_7 | TextureLS_07 | **TextureMS_07** | TextureHS_07 | **TextureNS_07** |
| Trial_8 | TextureLS_08 | **TextureMS_08** | TextureHS_08 | **TextureNS_08** |
| Trial_9 | TextureLS_09 | **TextureMS_09** | TextureHS_09 | **TextureNS_09** |
| Trial_10 | **TextureHS_10** | TextureNS_10 | **TextureLS_10** | TextureMS_10 |

restriction established in our first statistical analysis, that is, to include the first or second most salient objects in the first position. Considering that random orders provided by participants for each trial, it would be expected that 50% of the orders would fit our condition. Nonetheless, by the actual order given by each of the 40 participants in each of the 10 trials, a mean of 5.4 matches per participant was gotten. These 5.4 matches represent a 54% of all possible matches, which slightly exceeds the expected 50% value by random success.

The results in the second statistical analysis showed a degree of 50% agreements between the order based on our metric of saliency and based on the mode (Table 3(a) and (b)) respectively. It proves that our metric gives a good representation of the saliency perception in a prototypical person.

## 5   Conclusions

In this paper, we proposed a new method to measure the saliency for the texture that can be applied to 3D objects. Our measure is based in a saliency map by color. The proposal is characterized by: (1) To obtain the color of each pixel of an image in the RGB coordinate system, using the rendering of 3D graphics within a 2D image; (2) To transform the coordinates of the RGB color to the CIE Lab format and obtain a saliency value for each pixel; and (3) To calculate a salience value of the texture image. On the other hand, our metric of saliency can be applied to all kinds of visuals textures, considering that textures can be complex aggregations of shapes or colors, with variations or irregularities, whose contrast effects (light and shadow) have different intensity. Future works should explore other aspects contributing to the visual saliency of 3D objects other than their shape with respect to its vertices, color or size, and it analyze the influence of the context in which an object is placed, for a complete model of saliency for 3D objects.

## References

1. Huang, L., Pashler, H.: Quantifying object salience by equating distractor effects. Vis. Res. **45**, 1909–1920 (2005)
2. Hou, X., Zhang, L.: Saliency detection: a spectral residual approach. In: IEEE Conference on Computer Vision and Pattern Recognition 2007, Minneapolis, MN, pp. 1–8. IEEE (2007)
3. Itti, L.: Quantitative modelling of perceptual salience at human eye position. Vis. Cogn. **14**(4–8), 959–984 (2006)
4. Itti, L., Koch, C., Niebur, E.: A model of saliency-based visual attention for rapid scene analysis. IEEE. Trans. Pattern Anal. Mach. Intell. **20**(11), 1254–1259 (1998)
5. Li, C., Hamza, B.: A multiresolution descriptor for deformable 3D shape retrieval. Vis. Cogn. **29**(6–8), 513–514 (2013)
6. Raubal, M., Winter, S.: Enriching wayfinding instructions with local landmarks. In: International Conference on Geographic Information Science, Boulder, CO. Springer, Heidelberg, September 2002
7. Sampedro, M.J., et al.: Saliencia Perceptiva y Atención, In: La Atención (VI). Un enfoque pluridisciplinar, pp. 91–103 (2003)

8. Lahera, G., Freund, N., Sáin-Ruíz, J.: Asignación de relevancia (salience) y desregulación del sistema dopaminérgico. Elsevier Doyma. Revista de Psiquiatría y Salud Mental **6**(1), 45–51 (2013)
9. Gapp, K.-P.: Object localization: selection of optimal reference objects. In: Spatial Information Theory a Theoretical Basis for GIS, pp. 519–536 (1995)
10. Hoffman, D.D., Singh, M.: Salience of visual parts. Cognition **68**(1), 29–78 (1997)
11. Spotorno, S., Tatler, B.W., Faure, S.: Semantic consistency versus perceptual salience in visual scenes: findings from change detection. Acta Psycologica **142**, 168–176 (2013)
12. Stoia, L.: Noun phrase generation for situated dialogs. Ph.D. Thesis, Ohio State University, Ohio, pp. 1–179 (2007)
13. Gibson, J.J.: The perception of visual surfaces. Am. J. Psychol. **63**(3), 367–884 (1950)
14. Julesz, B.: Texture and visual perception. Sci. Am. **212**(2), 38–49 (1965)
15. Kimchi, R., Palmer, S.E.: Form and texture in hierarchically constructed patterns. J. Exp. Psychol. Hum. Percept. Perform. **8**(4), 521–535 (1982)
16. Stürzel, F., Spillmann, L.: Texture fading correlates with stimulus salience. Vis. Res. **41**(3), 2969–2977 (2001)
17. Landy, M.S., Graham, N.: Visual perception of texture. In: The Visual Neurosciences, p. 1106. MIT Press, Cambridge (2004)
18. Syeda-Mahmood, T.F.: Detecting perceptually salient texture regions in images. Comput. Vis. Image Underst. **76**(1), 93–108 (1999)
19. Terzic, K., Krishna, S., du Buf, J.M.H.: Texture features for object salience. Image Vis. Comput. **67**, 43–51 (2017)
20. Kadir, T., Brady, M.: Scale descriptors: texture description by salient scales. D.o.E.S. Robotics Research Laboratory, University of Oxford, Editor (2002)
21. Wang, Q., Yuan, Y., Yan P.: Visual saliency by selective contrast. IEEE Trans. Circuits Syst. Video Technol. **23**(7), 1150–1155 (2013)
22. Terzic, K., Krishna, S., du Buf, J.M.H.: A parametric spectral model for texture-based salience In: German Conference on Pattern Recognition, pp. 331–342. Springer, Cham, October 2015
23. Zhang, L., Yang, L., Luo, T.: Unified saliency detection model using color and texture features. PLoS ONE **11**(2), 1–14 (2016)
24. Lara, G., De Antonio, A., Peña, A.: A computational model of perceptual saliency for 3D objects in virtual environment (2017, in press)
25. Choungourian, A.: Color preferences and cultural variation. Percept. Mot. Skills **26** (3_suppl.), 1203–1206 (1968)
26. Abdelmounaime, S., Dong-Chen, H.: New brodatz-based image databases for grayscale color and multiband texture analysis. ISRN Mach. Vis. **2013**, 1–15 (2013)
27. Cimpoi, M.M., Subhransu Maji, et al.: Describing textures in the wild. In: Conference in Computer Vision and Pattern Recognition (CVPR), IEEE, June 2014
28. Lara, G., et al.: 3D objects shape relevance for saliency measure. In: 6th International Conference on Software Process Improvement (CIMPS 2017). México: Trends and Applications in Software Engineering. Springer (2017)

# Automating an Image Processing Chain of the Sentinel-2 Satellite

Rodrigo Rodriguez-Ramirez[1], María Guadalupe Sánchez[1],
Juan Pablo Rivera-Caicedo[2], Daniel Fajardo-Delgado[1],
and Himer Avila-George[3(✉)]

[1] TecNM, Instituto Tecnológico de Ciudad Guzmán,
Ciudad Guzmán, Jalisco, Mexico
{rodrigo12290785,msanchez,dfajardo}@itcg.edu.mx
[2] Cátedras CONACyT, Universidad Autónoma de Nayarit,
Tepic, Nayarit, Mexico
jprivera@uan.edu.mx
[3] Centro Universitario de los Valles, Universidad de Guadalajara,
Ameca, Jalisco, Mexico
himer.avila@academicos.udg.mx

**Abstract.** In this paper, a chain of satellite image processing using free software libraries is proposed, to estimate biophysical parameters using data from the Sentinel-2 satellite. In particular, the processing chain proposed allows atmospheric correction, resampling and spatial cropping of satellite images. To evaluate the functionality of the developed processing chain, the sugarcane cultivation of the Mexican region of Jalisco is introduced as a case study; from the selected scene, the leaf area index (LAI) is estimated using a model based on the Gaussian Process Regression technique, which is trained employing synthetic reflectance data created utilizing the PROSAIL radiative transfer model.

**Keywords:** Sentinel-2 · LAI · PROSAIL · Image processing chain
Gaussian process

## 1 Introduction

Mexico, like the rest of the world, is in a critical situation on issues of food security, due to the constant increase in population contrasted with a limited amount of land, water, and other natural resources [1]. For this reason, it is vitally important to develop new techniques to improve the use of natural resources and increase the production of food.

Currently, the use of remote sensing techniques through satellite images facilitates the elaboration of thematic maps that denote the state of agricultural and forestry resources, among others. With the processing of satellite images, soil conditions, vegetation types and their health can be discriminated. From these data, it is possible to obtain the cultivated area and even identify the plant species. Through the temporal analysis of satellite images, it is possible to monitor the evolution of different plant communities and crops.

© Springer Nature Switzerland AG 2019
J. Mejia et al. (Eds.): CIMPS 2018, AISC 865, pp. 216–224, 2019.
https://doi.org/10.1007/978-3-030-01171-0_20

Thanks to advances in space technology, new sensors have emerged on board space platforms which have improved the quality of the planet's sensed information. The Sentinel-2 is a satellite that incorporates the MSI (Multispectral Instrument) sensor capable of capturing images in 13 different bands where its bands stand out in the red-edge centered at 705, 740 and 783 nm [2], with this it is possible to carry out better studies of the plant cover of the planet; Additionally, with the information collected, predictions of the evolution of the crops can be made as well as the obtaining of characteristics of each one of them.

The development of methodologies that address forward and inverse problems has been essential in research in the fields of geomatics, physics, and remote sensing [3, 4]. These problems have been approached with empirical, statistical and machine learning models and complex physical-based models [5, 6], where the development of models that estimate accurately biophysical parameters from data coming from sensors remote areas has been a fundamental element in addressing these problems.

In recent years, several toolboxes have been developed focused on specific tasks such as downloading satellite images (SentinelSat), processing them (The Sentinel Application Platform, SNAP) and mapping biophysical variables (Automated Radiative Transfer Models Operator, ARTMO). However, it is necessary to orchestrate these isolated efforts to process the images, in this sense, it is needed to develop a chain of processing that allows the synergy between different libraries.

In this work, it is proposed to implement an image processing chain of the MSI sensor of the Sentinel-2 mission. The sugarcane cultivation of the Mexican region of Jalisco is presented as a case study to evaluate the functionality of the image processing chain.

## 2   Materials and Methods

### 2.1   Radiative Transfer Models

Radiative transfer models at the vegetation cover level simulate the complex interactions of electromagnetic energy with the different elements that make them up, e.g., leaves, stems and soil [7]. The complex configuration of the items that make up the covers has been simulated by different models designed in a specific way varying in degree of complexity when representing the crop [8].

To simulate the reflectance of sugar cane crops between 400 nm–2500 nm, the PROSAIL model was selected, which groups the models: PROSPECT [9] to simulate the leaves and 4SAIL [10] to simulate the geometry and structure of the crop canopy. This set of models has been widely used in different crops with structural characteristics like sugarcane [11] where the canopy is assumed as a turbid medium where the leaves are distributed randomly. The geometry of observation and lighting have been taken from the information derived from the images obtained from the Sentinel-2 satellite of the Copernicus program. The profile of the soil has been estimated by analyzing the images during the time of preparation of the land before sowing. Finally, the leaves inclination distribution function has been determined using an elliptical distribution of the mean angle of inclination.

Table 1 shows the parameterization of the PROSAIL model to be able to construct a reflectance database, which is composed of 2000 samples generated by the Latin Hypercube model [12]. For an extensive description of the parameters as well as the ranges of each parameter see [11]. Subsequently, the ARTMO program (Automated Radiative Transfer Models Operator) [13] was used to develop the database.

## 2.2 Statistical Model for Estimating the Leaf Area Index

Estimating biophysical variables from indirect measurements measured by sensors onboard space platforms are referred to as "investment problem"; The solution to this type of problem requires solving the ill-posed problem [14]. To do this different retrieval strategies have been evaluated through empirical solutions using inter-band relationships, numerical methods (used in conjunction with radiative transfer models) and statistical algorithms, and machine learning techniques [15]. In this work, the machine learning technique named Gaussian process regression (GPR) [3] has been used.

The GPR models provide a probabilistic approximation for regression problems with kernels [3], where the relationship between the spectral profiles and the biophysical parameter to be estimated is given by Eq. 1.

$$\hat{y} = f(x) = \sum_{i=1}^{N} \alpha_i K(x_i, x) + \alpha_0 \tag{1}$$

where $\{x_i\}_{i=1}^{N}$ are the spectral profiles used in the training, $\alpha_i \in R$ are the weights of each spectrum in the training database, $\alpha_0$ is the ordinate at the origin of the regression, and $K$ is the function that evaluates the similarity between the spectra of test and those of training.

In this work, the use of an RBF (Radial Basis Function) kernel of the type shown in Eq. 2 is proposed.

$$K(x_i, x_i) = v \exp\left(-\sum_{b=1}^{B} \frac{\left(x_i^b - x_j^b\right)^2}{2\sigma_b^2}\right) + \sigma_n^2 \delta_{xx'} \tag{2}$$

where $v$ is the scale factor, $B$ is the number of bands and the superscript $b$ in $x_i$ indicates the band $b$ of the profile $x_i$, $\sigma b$ is a factor that controls the propagation of each band, and $\sigma n$ is the standard deviation. The development of the GPR was carried out with the help of ARTMO.[1]

---

[1] ARTMO http://ipl.uv.es/artmo/.

**Table 1.** The range of values for the parameters for PROSAIL distributed with a Latin Hypercube model.

| Model | Symbol | Quantity | Unit | Min | Max |
|-------|--------|----------|------|-----|-----|
| PROSPECT | N | Leaf structure parameter | - | 1.5 | 2.5 |
| | Cab | Chlorophyll $a+b$ content | $\mu g\ cm^{-2}$ | 0 | 80 |
| | Cw | Equivalent water thickness | cm | 0.000 | 0.050 |
| | Cm | Dry matter content | $g\ cm^{-2}$ | 0.000 | 0.035 |
| 4SAIL | LAI | Leaf area index | - | 0 | 6 |
| | ALA | Average leaf angle | - | 30 | 60 |

## 2.3    Sentinel-2

The Sentinels are a new fleet of satellites explicitly designed to provide data and images for the Copernicus program of the European Commission. Sentinel-2 incorporates the MSI sensor carries an innovative high-resolution multispectral camera with 13 spectral bands that offer a new perspective of the Earth's surface and vegetation. The combination of high resolution and new spectral capabilities, as well as an encompassing field of view (290 km in width) and frequent overflights, provides unprecedented views of the earth.

Table 2 shows the resolution and wavelength of each of the 13 bands captured by the Sentinel-2 satellite.

## 2.4    Existing Tools

The following describes the existing tools for the processing of satellite images to obtain their biophysical parameters.

SentinelSat is an API that facilitates the download and recovery of Sentinel image metadata from the Copernicus Open Access Hub. To be able to make the downloads, the user needs to be registered in SciHub.[2]

For the conversion of the images in .jp2 format obtained from the Sentinel-2 satellite, RSGISLib[3] (Remote Sensing and GIS software library) was used, which is a set of tools for remote data processing and GIS datasets. Also, the OpenCV[4] and Numpy[5] libraries were used for matrix operations in the processing of images.

The NetCDF4 library is a tool developed in Python used to read and write files in NC format. NETCDF4 is one of the most important formats for storing and sharing scientific information besides facilitates the variables extraction and file values. This library added to the proposed processing chain for the model reading generated for ARTMO, and then it is used for the image processing.

---

[2] SciHub: https://scihub.copernicus.eu/twiki/do/view/SciHubWebPortal/APIHubDescription.
[3] RSGISLib: https://www.rsgislib.org/.
[4] OpenCV: https://opencv.org/.
[5] Numpy: http://www.numpy.org/.

**Table 2.** Resolution and wavelength of the 13 bands of the Sentinel-2 satellite.

| Band | Resolution (m) | Wavelength (nm) |
|------|----------------|-----------------|
| B01  | 60             | 443             |
| B02  | 10             | 490             |
| B03  | 10             | 560             |
| B04  | 10             | 665             |
| B05  | 20             | 705             |
| B06  | 20             | 740             |
| B07  | 20             | 783             |
| B08  | 10             | 842             |
| B08  | 20             | 865             |
| B09  | 60             | 945             |
| B10  | 60             | 1375            |
| B11  | 20             | 1610            |

The SEN2COR tool was used to perform the atmospheric correction of the images, this tool was developed externally to the SNAP/Sentinel Toolbox program and allows the processing of data from Sentinel-2 Level 1 to Level 2. As it is an external tool SNAP program, SEN2COR can be implemented in python independently.

## 3    Image Processing Chain

The satellite image processing chain consists of three modules: (a) Radiative transfer model, (b) Image acquisition and preprocessing module, and (c) Image processing module. The developed version of the processing chain in this work receives as input the model produced with the ARTMO software and the preprocessed image in .SAFE format of the SNAP software; as a result, it is obtained an image of the biophysical parameter that has been estimated. Next, the three modules that constitute the proposed image processing chain are described.

### 3.1    Radiative Transfer Retrieval Model

In this work, a module was developed as a complement to the ARTMO software; this module allows exporting the models developed in NetCDF format for the mapping of biophysical variables. Next, the PROSAIL model is used to create a database of 1000 reflectance samples according to the configuration in Table 1. The samples are processed using SimpleR, and subsequently, the model in NetCDF format is obtained. Figure 1 resumes the main processes linked to the development of the retrieval model.

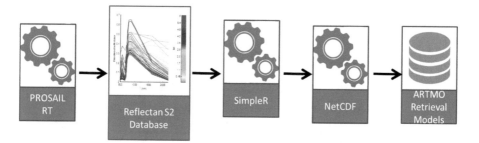

**Fig. 1.** The proposed scheme for the design of retrieval models using ARTMO.

### 3.2 Image Acquisition and Preprocessing

In Fig. 2, the method of acquisition and preprocessing of satellite data of the MSI sensor is described. The SentinelSat tool was used to search, download and retrieve the metadata of Sentinel satellite images from the Copernicus Open Access Hub. Then, the SNAP tool was used for the atmospheric correction and spatial resampling of the downloaded images.

### 3.3 Biophysical Mapping

Figure 3 shows the methodology followed to obtain the LAI parameter; which uses images in .jp2 format downloaded from the Sentinel 2 satellite platform. For this, the preprocessed images in Sect. 3.2 are converted to the GeoTIFF format using the RSGISLib tool; this is done with the purpose of facilitating the reading of these images in the next step. Then, the model generated in Sect. 3.1 is read, and the image is processed creating an image with LAI characteristics.

## 4  Case Study

To evaluate the processing chain described in Sect. 3, a Sentinel 2 sensor scene was downloaded on 05/28/2018; said scene corresponds to the municipality of Zapotiltic in Jalisco, Mexico (19° 38′0.35″ N, 103° 22′28.33″ W). Figure 4 shows the results, (a) the area selected in a combination of false color bands, and (b) the LAI map estimated from the development model with the PROSAIL radiative transfer model.

The LAI map estimated by the processing chain shows the congruence between the false color image that highlights the vigor of the vegetation with the red areas. The blue tones correspond to low LAI values related to fallow zones, scattered shrub vegetation and dry vegetation on the slopes. The algorithm estimates very well LAI areas close to zero for agricultural fields of bare ground and urban zones.

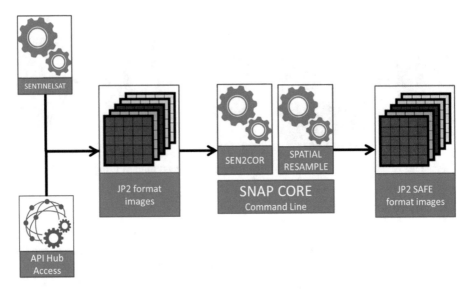

**Fig. 2.** The module for acquiring and preprocessing images of the Sentinel-2 satellite.

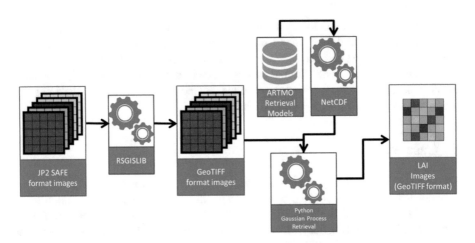

**Fig. 3.** The methodology for obtaining LAI images.

(a)                                                    (b)

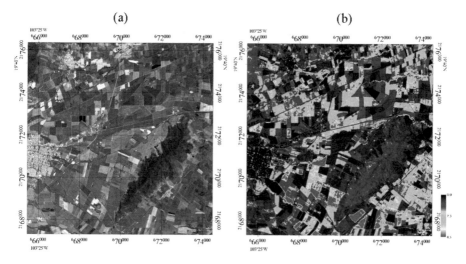

**Fig. 4.** (a) Image in a false color composite with the RGB combinations (865 nm, 665 nm, 560 nm); (b) LAI image estimated using the proposed processing chain.

## 5  Conclusions

This paper, presented a satellite image processing chain, implemented in Python, which allows cooperation with the ARTMO software, which contains a complete suite of algorithms that enable the estimation of biophysical parameters from reflectance data.

The results obtained show that the processing chain and the MSI Sensor have a lot of potential in the study of crops. The spatial variability in the images obtained after applying the chain of processing will allow sugarcane farmers to have tools that would enable improving the monitoring of said crop.

**Acknowledgments.** The first author thanks CONACYT for the scholarship granted to carry out your postgraduate studies. The other authors thank Dr. Jochem Verrelst from the Image Processing Laboratory of the University of Valencia, Spain, for giving access to the ARTMO tool.

## References

1. FAO: Anuario Estadístico de la FAO 2014 - La Alimentación y la Agricultura en América Latina y el Caribe. Organización de las Naciones Unidas para la Alimentación y la Agricultura, Santiago de Chile (2014)
2. European Space Agency: United space in Europe. https://www.esa.int/ESA
3. Svendsen, D.H., Martino, L., Campos-Taberner, M., García-Haro, F.J., Camps-Valls, G.: Joint Gaussian processes for biophysical parameter retrieval. IEEE Trans. Geosci. Remote Sens. **56**(3), 1718–1727 (2018)

4. Baret, F., Buis, S.: Estimating canopy characteristics from remote sensing observations: review of methods and associated problems. In: Advances in Land Remote Sensing: System, Modeling, Inversion and Application, pp. 173–201. Springer Netherlands, Amsterdam (2008)
5. Verrelst, J., Camps-Valls, G., Muñoz-Marí, J., Rivera, J.P., Veroustraete, F., Clevers, J.G., Moreno, J.: Optical remote sensing and the retrieval of terrestrial vegetation bio-geophysical properties – a review. ISPRS J. Photogramm. Remote Sens. **108**(1), 273–290 (2015)
6. Fernandes, R., Weiss, M., Camacho, F., Berthelot, B., Baret, F., Duca, R.: Development and assessment of leaf area index algorithms for the Sentinel-2 multispectral imager. In: IEEE Geoscience and Remote Sensing Symposium (2014)
7. Jacquemoud, S., Bacour, C., Poilve, H., Frangi, J.-P.: Comparison of four radiative transfer models to simulate plant canopies reflectance: direct and inverse mode. Remote. Sens. Environ. **74**(3), 471–481 (2000)
8. Liang, S.: Quantitative Remote Sensing of Land Surfaces. Wiley, Hoboken (2005)
9. Jacquemoud, S., Baret, F.: PROSPECT: a model of leaf optical properties spectra. Remote Sens. Environ. **34**(2), 75–91 (1990)
10. Verhoef, W., Bach, H.: Coupled soil–leaf-canopy and atmosphere radiative transfer modeling to simulate hyperspectral multi-angular surface reflectance and TOA radiance data. Remote Sens. Environ. **109**(2), 166–182 (2007)
11. Jacquemoud, S., Verhoef, W., Baret, F., Bacour, C., Zarco-Tejada, P.J., Asner, G.P., François, C., Ustin, S.L.: PROSPECT + SAIL models: a review of use for vegetation characterization. Remote Sens. Environ. **113**, S56–S66 (2009)
12. Stein, M.: Large sample properties of simulations using Latin hypercube sampling. Technometrics **29**(2), 143–151 (1987)
13. Verrelst, J., Rivera, J., Alonso, L., Moreno, J.: ARTMO: an automated radiative transfer models operator toolbox for automated retrieval of biophysical parameters through model inversion. In: EARSeL 7th SIG-Imaging Spectroscopy Workshop, Edinburgh, UK (2011)
14. Combal, B., Baret, F., Weiss, M., Trubuil, A., Macé, D., Pragnère, A., Myneni, R., Knyazikhin, Y., Wang, L.: Retrieval of canopy biophysical variables from bidirectional reflectance: using prior information to solve the ill-posed inverse problem. Remote Sens. Environ. **84**(1), 1–15 (2003)
15. Verrelst, J., Rivera, J.P., Veroustraete, F., Muñoz-Marí, J., Clevers, J.G., Camps-Valls, G., Moreno, J.: Experimental Sentinel-2 LAI estimation using parametric, non-parametric and physical retrieval methods – a comparison. ISPRS. J. Photogramm. Remote. Sens. **108**(1), 260–272 (2015)

# Automated Configuration of Monitoring Systems in an Immutable Infrastructure

Adrián Medina-González$^{(\boxtimes)}$, Sodel Vazquez-Reyes,
Perla Velasco-Elizondo, Huizilopoztli Luna-García,
and Alejandra García-Hernández

Autonomous University of Zacatecas, 98000 Zacatecas, ZAC, Mexico
{adrian.medina,vazquezs,pvelasco,hlugar,
alegarcia}@uaz.edu.mx

**Abstract.** Automated monitoring of Information Technology resources allows for the treatment of issues relating to availability, capacity, and other quality requirements. Currently, the use of monitoring systems in immutable infrastructures requires manually updating configurations every time a new server is launched, which is often time consuming and error prone. In this work, we propose a process to automate the configuration of monitoring systems in an immutable infrastructure. The process works for monitoring daemon services (Although Monit has the ability to monitor many other aspects of operating systems, this article only exposes the automation of the configuration of services or processes.) running in Debian-based operating systems and involves the use of technologies such as *Ansible*, *Monit*, and *Slack*. In contrast to manually updating configurations, the main advantages of the proposed method are: a reduction in time and user-friendliness when configuring the monitoring system.

**Keywords:** Immutable infrastructure · Configuration management
Service monitoring · *Ansible* · *Monit* · *Slack*

## 1 Introduction

Cloud computing allows users to access a variety of Information Technology (IT) resources through the network [1, 2]. Cloud computing has been adopted by many organizations because it is an easier, faster and cost-efficient alternative to deploying and running software systems without the need to purchase the underlying IT resources. Cloud computing services come in different forms, such as *Infrastructure as a Service* (IaaS), *Platform as a Service* (PaaS) and *Software as a Service* (SaaS) [3]. These forms of cloud computing allow users to treat IT resources as intangible and flexible, rather than physical and rigid [4].

An *immutable infrastructure* is a paradigm in which servers are never modified after they are deployed [5]. In contrast to traditional infrastructure paradigms, where servers are modified *in situ* by upgrading and downgrading packages, modifying configuration files, and deploying new code, immutable infrastructure entirely replaces outdated servers with new ones, rather than updating, or "mutating" them. The need for multiple mutations often results in complex configurations that make tasks such as

© Springer Nature Switzerland AG 2019
J. Mejia et al. (Eds.): CIMPS 2018, AISC 865, pp. 225–235, 2019.
https://doi.org/10.1007/978-3-030-01171-0_21

reproducing, replacing, scaling and recovering servers difficult [6]. In an immutable infrastructure, if servers need to be updated, fixed or modified, it is often easier, faster and more cost-efficient to create new servers with the required configurations from scratch. After validating them, the new servers are ready to use and the old ones are decommissioned. The main benefits of using an immutable infrastructure are the simplicity of implementation; reliability; stability; consistency; efficiency; and coherence. Together, these minimize the many points of conflict and failure of mutable infrastructures [6].

Service monitoring provides information, either in real time or for a specified time frame, which allows informed decision-making to prevent or correct failures and helps ensure that services provide their stated quality attributes, e.g. availability and capacity [1].

Today, the use of monitoring systems in immutable infrastructures often requires manually updating configurations every time a new server is launched. In Unix-based systems, the integration of a monitoring system into an immutable infrastructure is still under development and is not fully standardized. Consequently, such integration is not fully implemented or automated. Furthermore, updating configurations manually is often error prone and time consuming.

The main objective of this work is to propose a process for automating the configuration of monitoring systems in an immutable infrastructure, specifically, the process that automates the configuration of daemon services, via *Monit*,[1] that run on Debian-based operating systems and involve the use of technologies such as *Ansible*, *Monit* and *Slack*, as well as preconfigured and auto-generated configuration templates. In addition to contributing to the field of configuration management, and in contrast to manually updating configurations, the main advantages of the proposed method are: a reduction in time and user-friendliness when configuring the monitoring system.

This article is organized as follows; Sect. 2 explains the background concepts and technologies involved to in automatically supporting them. Next, Sect. 3 describes provides the details of how the background concepts and technologies are used in the proposed process. Section 4 presents a discussion and evaluation of this work. Finally, Sect. 5 includes a summary and outlines some avenues of future investigation.

## 2   Background

In this section, the background concepts of this work are explained, specifically, configuration management (CM) and service monitoring. We also provide some examples of technologies to automatically support them.

---

[1] Although Monit has the ability to monitor many other aspects of operating systems, this article only exposes the automation of the configuration of daemon services.

## 2.1    Configuration Management

Broadly, CM refers to the process of systematically handling changes to a system in a way that ensures the system maintains its integrity over time. Although this concept did not originate in the IT industry, it is now broadly used to refer to server CM.

Automation plays an essential role in server CM. It is the ideal mean to make a server reach a desirable state, previously not defined at all or defined by provisioning scripts written in a specific language and/or features of a CM tool. For servers, automation is, in fact, the heart of CM. There are many CM tools available on the market, each one with a specific set of scripting languages, features and different complexity levels. Popular tools include *Chef* [7], *Puppet* [8] and *Ansible* [9].

In this work, we use *Ansible* to handle CM. *Ansible* is an open source IT engine that automates CM, application deployment, cloud provisioning, intra-service orchestration, and many other IT tasks [10]. *Ansible* uses management and configuration scripts, called playbooks, to specify automation jobs on remote machines. Playbooks are written in a very simple scripting language called YAML [11], which allows users to describe them in a way that approaches plain English. *Ansible* uses a series of modules that can be run directly on remote hosts or through playbooks. These modules can control system resources, such as services, packages, and files, or manage the commands of the execution system [12].

## 2.2    Monitoring Systems

Service monitoring provides information, either in real time or for a specified time frame, which allows informed decision-making to prevent or correct failures and helps ensure that services provide their stated quality attributes, e.g. availability and capacity [1].

For CM, automation plays an essential role in service monitoring. There are a variety of monitoring system tools such as *Nagios* [13], *Icinga* [14], and *Monit* [15].

In this work, *Monit* has been selected to support service monitoring. *Monit* is a utility for managing and monitoring processes, programs, files, file systems, and directories on Unix-based systems [16]. *Monit* conducts automatic maintenance and repair and can execute meaningful causal actions in error situations, e.g., it can start a process if it does not run, restart a process if it does not respond, or stop a process if it uses too many resources.

## 3    The Proposed Process

In order to support the automated configuration of a monitoring system, we propose a three-step process that uses *Ansible*, *Monit*, and *Slack* to support CM, Monitoring, and Alerts respectively. Each will be performed by specific (sub)systems (Fig. 1). For simplicity, from now on daemon services will be referred simply as "services."

The implementation code for the process depicted in Fig. 1 can be downloaded from https://github.com/cracos/ansible-monit-slack. In the following sections, the details of this process are explained.

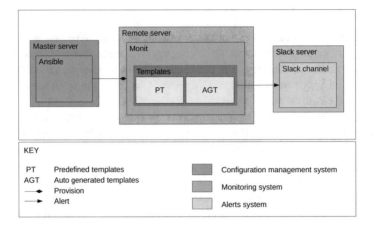

**Fig. 1.** Main elements of the proposed process to configure monitoring systems

## 3.1 Configuring a Monitoring System with *Ansible*

The first step in the proposed approach requires performing the CM of the monitoring system via *Ansible*, which is done using a variety of lists that contain information about the services to be (or not to be) monitored. In this work, two main types of lists are recognized: (i) user-provided lists; and (ii) auto-generated lists. Based on the information in these lists, *Ansible* will start to select or generate the corresponding configuration files. To reduce the probability of errors in configuration files, specific set operations are applied to the elements of these lists.

(i) User-provided lists:

- *Blacklist* (BL): *Ansible* configuration file where the user specifies the services that they do not wish to monitor.
- *Whitelist* (WL): *Ansible* configuration file where the user specifies the services that they do wish to monitor.

These lists can be generated in two main ways:

(a) Exclusion approach: requires the user to provide the BL. It is assumed that *Monit* will automatically monitor services that are running, except for those on the blacklist.
(b) Selective approach: requires the user to provide both the BL and WL.
(ii) Auto-generated lists using specific scripts:

- *List of predefined templates* (PT): generated from default CM information, i.e. port numbers.
- *List of services in execution* (SE): generated from active services that are running in the system.
- *Exclusion list by predefined templates* (ELPT): generated by performing the following operation (SE ∩ PT) − BL, as shown in Fig. 2.

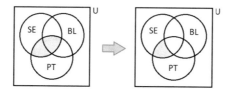

**Fig. 2.** Generation of the exclusion list using predefined templates

- *Exclusion list using basic templates (ELBT)*: generated by performing the following operation (SE − PT) − BL, as shown in Fig. 3.

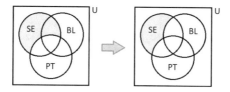

**Fig. 3.** Generation of the exclusion list using basic templates

- *Selective list using predefined templates (SLPT):* generated by performing the following operation {[(WL ∩ PT) ∩ SE] − BL}, as shown in Fig. 4.

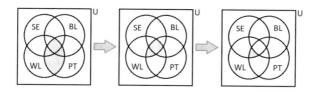

**Fig. 4.** Generation of the selective list using predefined templates

- *Selective list using basic templates (SLBT):* generated by performing the following operation {[(WL − PT) ∩ SE] − BL}, as shown in Fig. 5.

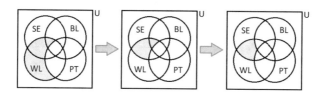

**Fig. 5.** Generation of the selective list using basic templates

## 3.2   Configuration of Services for *Monit*

The second step of the proposed approach is to perform the configuration in the Monitoring System, taking into consideration the configuration specified in *Ansible*. Service monitoring is performed using *Monit*.

*Monit* is configured by writing directives, a.k.a. rules, in a set of configuration files. The main configuration file is called *monitrc*, which consists of declarations of global definitions. In Debian-based systems this file is located in the `/etc/monit/` directory. *Monit* can use process-specific configuration files. These files are available, but not enabled, in the `/etc/monit/conf-available/` directory. To enable the desired configuration, symbolic links, a.k.a. symlinks, are created from the `/etc/monit/conf-enabled/` directory to the `conf-available` directory. This causes the configurations of the services to be monitored, added and loaded the next time *Monit* is started.

Predefined templates are stored in the `conf-available/preconfigured_templates/`[2] directory, while auto-generated templates are stored in `conf-available/autogenerated_templates/`. Auto-generated templates are stored for services that do not have a predefined template; when it is necessary to enable a service, only the corresponding symlink is created in the directory `conf-available`.

Configuration files can be structured in two different ways. The first uses the ".pid" file of the service to be monitored, which is a file that contains the process identification number (PID[3]). This is illustrated by the following example:

```
check process <unique name> with pidfile <path/to/pidfile.pid>
```

`<unique name>` is the unique name of the service to be monitored, and `<path/to/pidfile.pid>` is the absolute path where the service's .pid file is located. This form of configuration is recommended, as it defines the exact PID of the service of interest.

The alternative, which is used when the service to be monitored does not have a PID file, requires *Monit* to perform a process search using a regular expression or search pattern to find the process to be monitored. The following is an example:

```
check process <unique name> matching <regex>
```

`<regex>` is a regular expression or search pattern specifying the name of the service to be monitored. This form of configuration is only useful if the name of the process is unique. This alternative is used for generating basic templates. Each process has methods to start, stop, or restart its service that are used by *Monit* to execute an action on that service. For example:

---

[2] Templates are stored in a subdirectory to allow better organization.

[3] PID is an identification number that is automatically assigned to each process when it is created on Unix-like operating systems. Generally, it is stored in a ".pid" file, which allows other programs to find the PID of a running script.

```
check process <unique name> with pidfile <path/to/pidfile.pid>
start program = "/bin/xyz start"
stop program = "/bin/xyz stop"
restart program = "/bin/xyz restart"
```

Proactive monitoring can anticipate abnormal service events and take immediate action to prevent major incidents on the server. For example, if a server is consuming a lot of resources, *Monit* can stop or restart the server and send a notification. A simple example of how to tell *Monit* what to do, depending on the type of abnormal situation, is shown below:

```
check process <unique name> with pidfile <path/to/pidfile.pid>
start program = "/bin/xyz start"
stop program = "/bin/xyz stop"
if cpu > 80% for 4 cycles then alert
if cpu > 80% for 4 cycles then restart
```

In this example, if the process is consuming 80% or more processor resources for four *Monit* review cycles,[4] it will send an alert message and restart the process. Using this approach, a variety of checks can be enabled, such as detecting if a process is not running or detecting if the start of a service fails.

It is not always possible to have a predefined configuration template or enough information to generate it. For these situations, auto-generation of basic configuration templates was implemented. These templates contain information generated from a series of checks that are common for services. The checks are used to enable proactive monitoring of a service.

The logic of auto-generation of basic configuration templates is depicted in Fig. 6. First, the data of the services in the system is obtained based on the elements of the exclusion or selective lists, as seen in Fig. 6(a). Then, using the *Ansible* module "*Systemd*", information about running services is obtained, as in Fig. 6(b). From this information, the Control Group (Cgroup[5]) of the processes of interest is extracted. Then, from the Cgroup, the PID of the process is retrieved, shown in Fig. 6(c). Finally, using the ps[6] command, the information on how the process was launched is obtained, as in Fig. 6(d). This information is used to generate the search pattern that *Monit* will use to avoid the possibility of generating erroneous configurations with unwanted processes.

---

[4] By default, when *Monit* completes a programmed job, it goes to sleep for a configured period, then it wakes up and start monitoring again in an endless loop.

[5] Cgroup is a hierarchical grouping of processes managed by the OS kernel, and exposed through a special file system.

[6] The ps command is provides information about the processes that are running in the system.

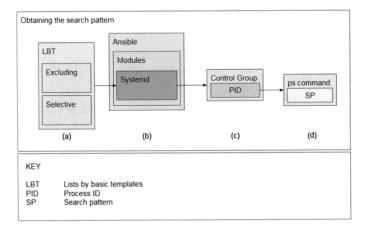

**Fig. 6.** The logic of auto-generation of basic configuration templates.

### 3.3 Notifying Alerts via *Slack*

The third step of the proposed approach requires defining an Alert System. By default, when something abnormal happens, *Monit* raises automatic alerts that are sent to the user through pre-configured emails. However, *Monit* can also be configured to send notifications to almost any messaging service [17]. In the method described in this article, *Slack* is used as a messaging service because it allows all notifications to be centralized and takes action for specific messages [18]. The configuration to generate alert notifications through *Slack* was included in each configuration template for *Monit* services, either preconfigured or self-generated, as described below.

Configuring *Slack* to send alerts requires using the `exec` command in the *Monit* configuration files. The following example is a variant of one presented above:

```
check process <unique name> with pidfile <path/to/pidfile>
start program = "/bin/xyz start"
stop program = "/bin/xyz stop"
if cpu > 80% for 4 cycles then exec "/etc/monit/SlackAlert.sh"
```

Here, the configuration specifies that whenever the processor goes over 80% usage, a notification will be sent to a *Slack* channel. *Incoming WebHooks*[7] are used in this work. A Webhook provides a unique URL to which the message text is sent [19]. Using this method, *Monit* alerts can be received in a specific *Slack* chat room.

### 3.4 Generating Logs

When the monitoring system is deployed successfully, a log file is generated containing a summary of all the services and how they were configured. The log file can be found in the path `/etc/monit/rs.log`.

---

[7] They are a simple way to post messages from applications in Slack.

# 4   Discussion and Evaluation

There are four main aspects from which we discuss and evaluate the proposed approach: (i) CM tools; (ii) Monitoring tools; (iii) Messaging tools and (iv) Time and error reduction when configuring the monitoring system.

## 4.1   CM Tools

As mentioned previously, there are many CM tools available in the market: popular choices include *Chef* [7], *Puppet* [8], and *Ansible* [9]. We chose *Ansible* as our CM because it compatible with, and provides an easy way to configure, *Monit*. *Ansible* CM playbook syntax is built on top of YAML, which is a scripting language designed to be easy for humans to read and write. That is, it promotes user-friendliness. Also *Ansible* is agentless, which means there is no need to pre-install an agent or any other software on hosts (i.e. remote servers). It requires a server to have the SSH utility and Python 2.5 or later installed [20].

## 4.2   Service Monitoring Tools

Today, widely-used tools for monitoring services include *Nagios* [13], *Icinga* [14], and *Monit* [15]. A real benefit of *Monit* is its easy configuration and syntax. While developers have to dig through a bunch of files in order to create a simple check in *Nagios*, *Monit* allows them to simply use one (human-readable) configuration file in the correct path for it o work. *Nagios* and *Icinga* are not user-friendly to new users and involve complicated configurations because their installation and configuration require extensive knowledge of the underlying operating system.

## 4.3   Messaging Tools

The most important reason people use the *Slack* instant messaging tool is that it can be easily integrated with tools such as *Trello* [21], *GitHub* [22], *Dropbox* [23], *MailChimp* [24], and many more. Furthermore, *Slack* allows centralized events integrated into chat rooms.

## 4.4   Time Reduction

As we will discuss in Sect. 5, we are planning to perform a systematic study to quantitatively evaluate the reduction in time and error of the proposed approach. However, we have performed some experiments and the results seem very promising. We performed the described process to deploy instances of the monitoring system dealing with one to twenty two services and it took from 9 to 28 min to do it, respectively. When common changes to the existing configurations of these instances were performed (e.g. changes in port numbers, services' files dependencies, checks) it took from 9 to 13 min when dealing with one to twenty two services.

## 4.5    Other Thoughts

There is a repository on *GitHub* (https://github.com/pgolm/ansible-role-monit) with an *Ansible-Monit* configuration system. This system is similar to the one proposed in this paper but does not implement the auto-configuration of services and integration of alert notifications with instant messaging tools as ours does.

## 5    Summary and Future Work

The use of monitoring systems in immutable infrastructures sometimes requires manually updating configurations every time a new server is launched, which is often time consuming and error prone. In this paper, we described a process to automate the configuration of monitoring systems in an immutable infrastructure. The process works for monitoring daemon services running in Debian-based operating systems and involves the use of technologies such as *Ansible, Monit*, and *Slack*.

Future work in this area includes performing a systematic study to validate the benefits of the process presented in this paper. The study will be performed in two IT companies in the state of Zacatecas, Mexico: Sharing Economy Tools, which is a Mexican company that offers software development services, and Tinkerware, which is a Mexican company that provides infrastructure automation solutions and services to software development companies.

In this effort we worked with daemon services in Debian-based operating systems. We plan to explore the feasibility to automate monitoring of services that are not daemon. We plan also explore the use of these proposed approach with non-Debian-based operating systems. In addition, we are considering incorporating more of the monitoring options offered by *Monit* into our process, such as *host*, or *filesystems*.

**Acknowledgments.** The authors would like to extend thanks to Tinkerware and Sharing Economy Tools, and especially to Alfonso Álvarez Sánchez and Agustín Rumayor Barraza, members of each of these companies, who provided insight and expertise that greatly assisted the process presented in this paper.

## References

1. Fatema, K., Emeakaroha, V.C., Healy, P.D., Morrison, J.P., Lynn, T.: A survey of cloud monitoring tools: taxonomy, capabilities and objectives. J. Parallel Distrib. Comput. **74**(10), 2918–2933 (2014)
2. Youseff, L., Butrico, M., Da Silva, D.: Toward a unified ontology of cloud computing. In: Grid Computing Environments Workshop, GCE 2008 (2008)
3. Syed, H.J., Gani, A., Ahmad, R.W., Khan, M.K., Ahmed, A.I.A.: Cloud monitoring: a review, taxonomy, and open research issues. J. Netw. Comput. Appl. **98**, 11–26 (2017)
4. Fowler, M.: The disposable infrastructure // Speaker deck (2017). https://speakerdeck.com/mlfowler/the-disposable-infrastructure. Accessed 25 Nov 2017
5. Stella, J.: An introduction to immutable infrastructure - O'Reilly media (2015). https://www.oreilly.com/ideas/an-introduction-to-immutable-infrastructure. Accessed 18 June 2018

6. Virdó, H., DigitalOcean: Immutable infrastructure | DigitalOcean (2017). https://www.digitalocean.com/community/tutorials/what-is-immutable-infrastructure. Accessed 12 Apr 2018
7. Chef Software Inc.: Chef - Automate IT infrastructure (2018). https://www.chef.io/chef/. Accessed 20 June 2018
8. Puppet Enterprise: Deliver better software, faster (2018). https://puppet.com/. Accessed 20 June 2018
9. Red Hat Inc.: Ansible (2018). https://www.ansible.com/. Accessed 26 Nov 2017
10. Red Hat Inc.: How Ansible works | Ansible.com (2018). https://www.ansible.com/overview/how-ansible-works. Accessed 20 Mar 2018
11. YAML: The official YAML Web Site (2018). http://yaml.org/. Accessed 20 June 2018
12. Red Hat Inc.: Working with modules—Ansible documentation (2018). https://docs.ansible.com/ansible/latest/user_guide/modules.html. Accessed 17 June 2018
13. Nagios Enterprises LLC.: Nagios - Network, server and log monitoring software (2018). https://www.nagios.com/. Accessed 20 June 2018
14. Icinga Open Source Monitoring: Icinga – Open source monitoring (2018). https://www.icinga.com/. Accessed 20 June 2018
15. Tildeslash Ltd.: Monit (2017). https://mmonit.com/monit/. Accessed 16 Mar 2017
16. Tildeslash Ltd.: Monit manual (2018). https://mmonit.com/monit/documentation/monit.html. Accessed 20 Mar 2018
17. Tildeslash Ltd.: M/Monit | Slack Notification (2018). https://mmonit.com/wiki/MMonit/SlackNotification. Accessed 17 June 2018
18. Slack Technologies: What is slack? – Slack help center (2018). https://get.slack.help/hc/en-us/articles/115004071768-What-is-Slack. Accessed 20 Mar 2018
19. Slack Technologies: Incoming webhooks (2018). https://api.slack.com/incoming-webhooks. Accessed 17 June 2018
20. Hochstein, R.M.L.: Ansible: Up and Running, 2nd edn. O'Reilly, Newton (2017)
21. Trello. https://trello.com/. Accessed 20 June 2018
22. GitHub. https://github.com/. Accessed 20 June 2018
23. Dropbox. https://www.dropbox.com/. Accessed 20 June 2018
24. MailChimp: Marketing platform for small businesses (2018). https://mailchimp.com/. Accessed 20 June 2018

# Using Convolutional Neural Networks to Recognition of Dolphin Images

Yadira Quiñonez[(✉)], Oscar Zatarain, Carmen Lizarraga,
and Juan Peraza

Universidad Autónoma de Sinaloa, Facultad de Informática Mazatlán,
Av. Universidad y Leonismo Internacional S/N, Mazatlán, Mexico
{yadiraqui,ozatarain,carmen.lizarraga,
jfperaza}@uas.edu.mx

**Abstract.** Classification of specific objects through Convolutional Neural Networks (CNN) has become an interesting research line in the area from information processing and machine learning, main idea is training a image dataset to perform the classifying a given pattern. In this work, a new dataset with 2504 images was introduced, the method used to train the networks was transfer learning to recognition of dolphin images. For this purpose, two models were used: Inception V3 and Inception ResNet V2 to train on TensorFlow platform with different images, corresponding to the four main classes: dolphin, dolphin_pod, open_sea, and seabirds. The paper ends with a critical discussion of the experimental results.

**Keywords:** Convolutional neural networks · Machine learning
Deep learning · TensorFlow · Inception V3 · Inception ResNet V2

## 1 Introduction

Machine Learning (ML) is considered as one of the most important area of artificial intelligence, it has had a surprising growth and has been applied in different areas, such as pattern recognition, natural language processing and computational learning [1], it is a field in continuous evolution that is based on statistical techniques. Some of the typical problems that can be found are the recognition of personal features [2, 3], detection of faces [4–6], recognition of types of landscapes [7], recognition of handwritten digits [8, 9], among others.

Deep learning is a subfield of machine learning, which has made important advances through computational models formed by multiple layers of processing [10]. It has become an increasingly extensive and interesting research object due to several factors, mainly, in the exponential growth in the volume of information, in the advances in technologies, such as: more powerful and economic data processing units, and, large capacity storage units, as a result of this, it has been possible to develop algorithms and models to carry out complex operations [11]. Currently, it is widely used for different applications in Artificial Intelligence (AI) including computer vision [12, 13], speech recognition [14], robotics [15], semantic parsing [16], transfer learning [17], natural language processing [18, 19], among others.

© Springer Nature Switzerland AG 2019
J. Mejia et al. (Eds.): CIMPS 2018, AISC 865, pp. 236–245, 2019.
https://doi.org/10.1007/978-3-030-01171-0_22

## 2   Formal Description of the Problem

Since ancient times, the fishing industry has tried to know exactly the location of shoals fish. In recent years, significant progress has been made in the incorporation of technology to locate shoals of fish, where use advanced techniques to carry out the extraction of a greater number of catches safely, quickly, and with less operational costs. However, the main confirmation technique for the detection of shoals of fish is carried out from a helicopter; basically, it consists of detecting breeze patterns, agglomerations of species such as seabirds and dolphins.

In this sense, we propose a transfer learning approach with convolutional neural networks to recognition of dolphin images. For this purpose, two models were used: Inception V3 and Inception ResNet V2 to train different images corresponding to the four main classes (see Fig. 1): dolphin, dolphin_pod, open_sea, and seabirds.

(a) Dolphin          (b) Dolphin_Pod          (c) Open_Sea          (d) Sea Birds

**Fig. 1.** Examples of images to train the Inception V3 and Inception ResNet V2 models.

Images data set used consists of different images that represent the desired objectives, such as the dolphins, seabirds, and the open sea. A search was made in different sources to obtain an adequate database and it was not possible to find a database with the representative characteristics of the model. In this sense, a dataset was created with images similar to the needs of the model, in order to reduce the margin of error and increase the effectiveness of the model in real conditions, these images were obtained from different videos taken from the air in open sea, applying some rotation and lighting techniques to increase the database. To carry out the training of the models, the data were prepared, first, the information was collected, then the information was cleaned and the information was separated. At the end, we obtained a dataset with 2504 images that were randomly divided with 80% for the training process, 10% for validation, and 10% for the test.

The analysis and classification of images is carried out by *Tensorflow*; it is a powerful library of artificial intelligence developed and supported by Google [20]. Tensorflow is designed to perform complex numerical computations, due to its flexible architecture allows its implementation in a variety of platforms such as: CPUs, GPUs, and TPUs, which allows developers to experiment with new optimizations and training algorithms.

# 3  Theoretical Foundation

Artificial Neural Networks (ANN) have been studied since the 60's until today by various researchers in the scientific community to solve problems in many different application areas. It is mainly based on the functioning of the human brain, they are dynamic auto-adaptive systems, which adjust to the elements that composed the system, are able to adapt to the new conditions of the system. Basically, the ANN's as its name indicates, are composed for a number of interconnected neurons, where its input parameters are the set of signals received from the environment, after, calculations are performed using an activation function [21]. That is, with this type of systems it is possible to process information in real time to create artificial models that solve difficult problems, such as: the prediction of time series, data mining, classification problems, pattern recognition, adaptive control, among others.

In recent years, CNN's have received substantial attention and have been very studied by different researchers mainly from information processing and machine learning community [22], and it has become very popular due to thanks to the high-performance GPUs or CPU clouds. CNN's are the current architecture for related tasks to the image classification, image segmentation, object recognition, and face recognition. These network assume explicitly that the input data are images, which allows encoding certain properties in the architecture. This makes the forwarding function more efficient and significantly reduces the number of parameters required by the network. Therefore, the computational complexity of the CNN's is determined by the convolutional layers, and the number of parameters is related to the fully connected layers [23].

## 3.1  Architecture

Regularly, a CNN's consist of a stack of convolutional modules that carry out feature extraction, they are usually composed by a set of layers that can be grouped by their functionalities. Three types of layer are used: Convolutional Layers, Pooling Layers and Classification Layers (see Fig. 2).

**Convolutional Layers:** Basically, in the convolution step the goal is to extract features from the input image, considering each image as a matrix of pixel values, then, a matrix called a *filter* or *kernel* or *feature detector* is obtained to slide the filter over the image and computing the dot product, this procedure is called Convolved Feature or Feature Map. The size of the convolved map is controlled by three parameters: Depth (number of filters used for the convolution operation), Stride (number of pixels to slide the filter matrix over the input matrix), and Zero-padding (it is recommended to pad the input matrix with zeros around the border to apply the filter at the edges of the input image matrix). Additionally, every time the convolution operation is performed, at the same time, the operation called ReLU is carried out, which consists in replacing all negative pixel values in the feature map by zero [24, 25].

**Pooling Layers:** It is common to periodically insert a pooling layer between successive convolutional layers. Its main function is to progressively reduce the spatial size of the data of the images that are extracted by the convolutional layers to reduce

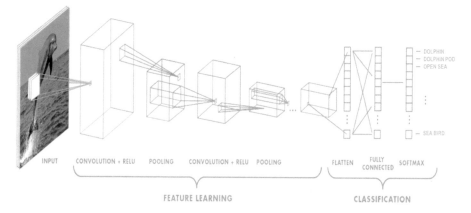

**Fig. 2.** Convolutional neural network architecture and the training process is composed for several blocks: convolution layers + ReLU, pooling layers, and fully connected.

the amount of parameters and calculations in the network, it is also used to control overfitting. This layer operates independently in each depth segment of the input data and reduces it spatially, it can be used different types of pooling: Max, Average, Sum. Max Pooling extracts the largest element from map, stores its maximum value and discards the other values. It can also take the average or the sum of all the elements from map, however, max pooling is the algorithm that is commonly used in the CNN's [24, 25].

**Fully Connected Layer:** In this layer, the classification of the features that have been extracted by the convolutional layers and reduced by the pooling layers is carried out. In the fully connected layer, each node of the layer is connected to each node in the layer that precedes it. That is, in this layer the Multi Layer Perceptron and a softmax activation function in the output layer are used, with the aim of using these features to classify the input image into several classes according to the training dataset [24, 25].

## 3.2   Transfer Learning Approach

Transfer learning is a technique that greatly shortens the process by taking a piece from a model that has been previously trained in a specific task and reuses it in a new model. According to Pan and Yang [26], "transfer learning techniques try to transfer the knowledge from some previous tasks to a target task when the latter has fewer high-quality training data". This means, that it is possible to use the knowledge that a model has learned in a specific task from a large amount of training data, in another new task where there is little objective training data [27].

## 3.3   Models

Inception V3 model is a deep convolutional neural network, and has been designed to perform computational calculations in an efficient way using 12 times less parameters than other models, this allows it to be implemented in less powerful systems [28].

Inception ResNet V2 was born from the implementation of a more simplified and optimized architecture of Inception V3 and the use of residual connections. Residual connections were introduced by He et al. [29], basically, these connections allow shortcuts in the model for training very deep architectures.

## 4 Experimental Results

As previously mentioned, one of the possible signs that indicate the presence of a shoals of fish is when there is an agglomeration of dolphins and different species of birds at a specific point. In this sense, four sets of data have been used to train the models, it has taking into account 2504 different images corresponding to the four main classes: Dolphin (626), Dolphin_Pod (626), Open_Sea (626), and Sea Birds (626).

In the first phase, the training tests were carried out in both models: Inception V3 and Inception ResNet V2, using the training algorithm with the same parameters; algorithm parameters for both models are listed below:

- how_many_training_steps = 4000
- learning_rate = 0.01
- testing_percentage = 10
- validation_percentage = 10
- eval_step_interval = 10
- train_batch_size = 100
- test_batch_size = 1
- validation_batch_size = 100
- flip_left_right = false.

In Figs. 3 and 4 respectively, the results obtained in each of the models are shown in relation to the accuracy of training and validation.

The next phase, was to process an isolated data set (which is not used in the training process) to obtain the predicted label, and accuracy for each model. Table 1 shows the results obtained in relation to prediction and accuracy of both model.

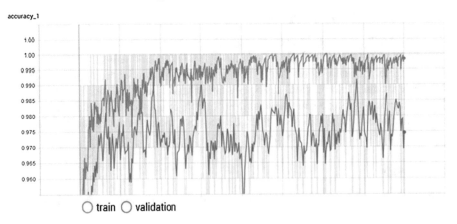

**Fig. 3.** Accuracy of the training and validation in the Inception V3 model.

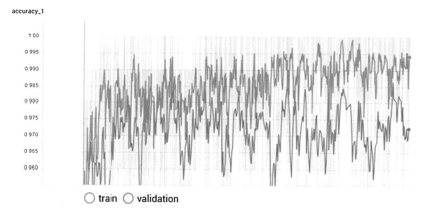

**Fig. 4.** Accuracy of the training and validation in the Inception ResNet V2 model.

**Table 1.** Results of the isolated data set with the prediction and precision of each model.

| Image | Label | Prediction | | Accuracy | |
|---|---|---|---|---|---|
| | | Inception V3 | ResNet V2 | Inception V3 | ResNet V2 |
| d1 | dolphin | dolphin | dolphin | 0.999867 | 0.999201 |
| d2 | dolphin | dolphin | dolphin | 0.999197 | 0.9992054 |
| d3 | dolphin | dolphin | dolphin | 0.999290 | 0.99976665 |
| d4 | dolphin | dolphin | dolphin | 0.999112 | 0.99902546 |
| d5 | dolphin | dolphin | dolphin | 0.999584 | 0.99907076 |
| d6 | dolphin | dolphin | dolphin | 0.993619 | 0.99982435 |
| d7 | dolphin | dolphin | dolphin | 0.999377 | 0.99839336 |
| d8 | dolphin | dolphin | dolphin | 0.993750 | 0.99051875 |
| d9 | dolphin | dolphin | dolphin | 0.996684 | 0.99861753 |
| d10 | dolphin | dolphin | dolphin | 0.998350 | 0.98963416 |
| dp1 | dolphin_pod | dolphin_pod | dolphin_pod | 0.999894 | 0.9999219 |
| dp2 | dolphin_pod | dolphin_pod | dolphin | 0.606840 | 0.8279542 |
| dp3 | dolphin_pod | dolphin_pod | dolphin_pod | 0.997866 | 0.99991286 |
| dp4 | dolphin_pod | dolphin_pod | dolphin_pod | 0.950985 | 0.9474429 |
| dp5 | dolphin_pod | dolphin_pod | dolphin_pod | 0.997667 | 0.98476464 |
| dp6 | dolphin_pod | open_sea | open_sea | 0.899063 | 0.68378663 |
| dp7 | dolphin_pod | dolphin | dolphin | 0.733563 | 0.96086633 |
| dp8 | dolphin_pod | dolphin | dolphin | 0.823670 | 0.97589886 |
| dp9 | dolphin_pod | dolphin | dolphin | 0.989146 | 0.9984806 |
| dp10 | dolphin_pod | dolphin | dolphin | 0.995447 | 0.99773127 |
| os1 | open_sea | open_sea | open_sea | 0.999750 | 0.9993788 |
| os2 | open_sea | open_sea | open_sea | 0.708610 | 0.7682051 |
| os3 | open_sea | open_sea | open_sea | 0.915449 | 0.5169322 |

*(continued)*

**Table 1.** (*continued*)

| Image | Label | Prediction | | Accuracy | |
|-------|-------|------------|-----------|----------|-----------|
|       |       | Inception V3 | ResNet V2 | Inception V3 | ResNet V2 |
| os4   | open_sea | open_sea | open_sea | 0.904816 | 0.8232759 |
| os5   | open_sea | open_sea | open_sea | 0.974195 | 0.933742 |
| os6   | open_sea | open_sea | open_sea | 0.988578 | 0.992273 |
| os7   | open_sea | open_sea | open_sea | 0.997968 | 0.9991886 |
| os8   | open_sea | open_sea | dolphin | 0.759634 | 0.3859054 |
| os9   | open_sea | open_sea | open_sea | 0.991958 | 0.8676522 |
| os10  | open_sea | open_sea | open_sea | 0.952734 | 0.7189585 |
| sb1   | seabird | dolphin_pod | dolphin_pod | 0.793808 | 0.72338843 |
| sb2   | seabird | seabird | seabird | 0.970322 | 0.9957217 |
| sb3   | seabird | seabird | seabird | 0.992792 | 0.9839759 |
| sb4   | seabird | seabird | seabird | 0.679836 | 0.6764687 |
| sb5   | seabird | open_sea | open_sea | 0.998229 | 0.9910097 |
| sb6   | seabird | dolphin_pod | seabird | 0.811937 | 0.81788874 |
| sb7   | seabird | seabird | seabird | 0.972137 | 0.93101656 |
| sb8   | seabird | seabird | seabird | 0.998681 | 0.9925035 |
| sb9   | seabird | seabird | seabird | 0.997083 | 0.9676014 |
| sb10  | seabird | seabird | Seabird | 0.515455 | 0.89151603 |

Tables 2 and 3 show the confusion matrices generated from the predictions of the isolated data set for each model. In general, both models get a good performance in relation to the class of dolphins and open sea, in relation to the Dolphin Pod class, both models make a good classification, however, sometimes it is classified as dolphin instead of Dolphin Pod.

**Table 2.** Confusion matrix of Inception V3.

|             | Predictions | | | |
|-------------|---------|-------------|----------|---------|
|             | dolphin | dolphin_pod | open_sea | seabird |
| dolphin     | **10**  | 0           | 0        | 0       |
| dolphin_pod | 4       | **5**       | 1        | 0       |
| open_sea    | 0       | 0           | **10**   | 0       |
| seabird     | 0       | 2           | 1        | **7**   |

**Table 3.** Confusion matrix of Inception ResNet V2.

|             | Predictions | | | |
|-------------|---------|-------------|----------|---------|
|             | dolphin | dolphin_pod | open_sea | seabird |
| dolphin     | **10**  | 0           | 0        | 0       |
| dolphin_pod | 5       | **4**       | 1        | 0       |
| open_sea    | 1       | 0           | **9**    | 0       |
| seabird     | 0       | 1           | 1        | **8**   |

**Table 4.** Accuracy isolated data set.

|  | dolphin | dolphin_pod | open_sea | Seabird |
|---|---|---|---|---|
| Inception V3 | 0.997883 | **0.910650** | 0.919369 | 0.875187 |
| Inception ResNet V2 | 0.997326 | *0.983011* | *0.846623* | **0.907087** |

Table 4 shows the results of the accuracies obtained in the predictions by class, it can be seen that Inception ResNet V2 model throws outstanding values in the class *dolphin_pod*, however, according to the result of confusion matrix, this model only hit four times out of ten, this means that, the precision can not be considered with respect to the other model that managed to correctly label five images; due to this, in the later analyses this model will be excluded.

## 5    Conclusions and Further Research Work

In this paper, we have analyzed two models: Inception V3 and Inception ResNet V2 to train on TensorFlow platform 2504 different images, corresponding to the four main classes: Dolphin (626), Dolphin_Pod (626), Open_Sea (626). According to the results the Inception V3 model presents better results and a high performance in the training (99.35%), validation (97.34%), and testing (99.60%) phases. It can be said that Inception V3 is the one that best adapts and offers an optimal and stable precision in the general behavior of the system.

It has been verified that the algorithm is not affected by changes of scale and image rotation due to preprocessing performed, and at the same time, it requires a smaller amount of processing and offers the possibility to optimize the model, this makes it a viable option even to develop it in a mobile application.

It would be interesting to evaluate and test the performance in convolutional neural networks with other models such as: MobileNet V2, NASNet-A (Large), and PNASNet-5 (Large), and Xception in order to obtain a better performance. And at the same time, experiment with other deep learning platforms such as: Caffe or Pytorch to verify the results In addition, It would be interesting to try the best model in Unmanned Aerial Vehicle (UAV) to test the performance of the model in a real scenario.

**Acknowledgments.** The authors would like to thank Universidad Autónoma de Sinaloa for supporting and financing this research project.

## References

1. Liu, W., Wang, Z., Liu, X., Zeng, N., Liu, Y., Alsaadi, F.E.: A survey of deep neural network architectures and their applications. Neurocomputing **234**, 11–26 (2017)
2. Huang, F., Sun, T., Bu, F.: Generation of person-specific 3D model based on single photograph. In: 2nd IEEE International Conference on Computer and Communications, pp. 704–707. IEEE Press (2016)

3. Choi, W., Chao, Y.W., Pantofaru, C., Savarese, S.: Discovering groups of people in images. In: Fleet, D., Pajdla, T., Schiele, B., Tuytelaars, T. (eds.) Computer Vision – ECCV 2014. Lecture Notes in Computer Science, vol. 8692, pp. 417–433. Springer, Cham (2014)

4. Ouarda, W., Trichili, H., Alimi, A.M., Solaiman, B.: Face recognition based on geometric features using support vector machines. In: 6th International Conference of Soft Computing an Pattern Recognition. pp. 89–95 (2014)

5. Chen, Q., Kotani, K., Lee, F.: Face recognition using multiple histogram features in spatial and frequency domains. In: 12th International Conference on Signal-Image Technology Internet-Based Systems, pp. 204–208. IEEE Press (2016)

6. Atallah, R.R., Kamsin, A., Ismail, M.A., Abdelrahman, S.A., Zerdoumi, S.: Face recognition and age estimation implications of changes in facial features: a critical review study. IEEE Access 6, 28290–28304 (2018)

7. Bradbury, G., Mitchell, K., Weyrich, T.: Multi-spectral material classification in landscape scenes using commodity hardware. In: Wilson, R., Hancock, E., Bors, A., Smith, W. (eds.) Computer Analysis of Images and Patterns. Lecture Notes in Computer Science, vol. 8048, pp. 209–216. Springer, Berlin (2013)

8. Lu, W.S.: Handwritten digits recognition using PCA of histogram of oriented gradient. In: IEEE Pacific Rim Conference on Communications, Computers and Signal Processing, pp. 1–5. IEEE Press (2017)

9. Larasati, R., KeungLam, H.: Handwritten digits recognition using ensemble neural networks and ensemble decision tree. In: International Conference on Smart Cities, Automation Intelligent Computing Systems, pp. 99–104. IEEE Press (2017)

10. LeCun, Y., Bengio, Y., Hinton, G.: Deep learning. Nature 521, 436–444 (2015)

11. Awad, M., Khanna, R.: Deep learning. In: Efficient Learning Machines, pp 167–184. Apress, Berkeley (2015)

12. Wu, Q., Liu, Y., Li, Q., Jin, S., Li, F.: The application of deep learning in computer vision. In: Chinese Automation Congress, pp. 6522–6527. IEEE Press (2017)

13. Goswami, T.: Impact of deep learning in image processing and computer vision. In: Anguera, J., Satapathy, S., Bhateja, V., Sunitha, K. (eds.) Microelectronics, Electromagnetics and Telecommunications. Lecture Notes in Electrical Engineering, vol. 471, pp. 475–485. Springer, Singapore (2018)

14. Sustika, R., Yuliani, A.R., Zaenudin, E., Pardede, H.F.: On comparison of deep learning architectures for distant speech recognition. In: 2nd International Conferences on Information Technology, Information Systems and Electrical Engineering, pp. 17–21. IEEE Press (2017)

15. Miyajima, R.: Deep learning triggers a new era in industrial robotics. MultiMedia 24(4), 91–96 (2017)

16. Heck, L., Huang, H.: Deep learning of knowledge graph embeddings for semantic parsing of Twitter dialogs. In: Global Conference on Signal and Information Processing, pp. 597–601. IEEE Press (2014)

17. Moriya, S., Shibata, C.: Transfer learning method for very deep CNN for text classification and methods for its evaluation. In: 42nd Annual Computer Software and Applications Conference, pp. 153–158 (2018)

18. Alshahrani, S., Kapetanios, E.: Are deep learning approaches suitable for natural language processing? In: Métais, E., Meziane, F., Saraee, M., Sugumaran, V., Vadera, S. (eds.) Natural Language Processing and Information Systems. Lecture Notes in Computer Science, vol. 9612, pp. 343–349. Springer, Cham (2016)

19. He, X., Deng, L.: Deep learning in natural language generation from images. In: Deng, L., Liu, Y. (eds.) Deep Learning in Natural Language Processing, pp. 289–307. Springer, Singapore (2018)

20. TensorFlow. https://www.tensorflow.org/
21. Guresen, E., Kayakutlu, G.: Definition of artificial neural networks with comparison to other networks. Procedia Comput. Sci. **3**, 426–433 (2011)
22. Gu, J., Wang, Z., Kuen, J., Ma, L., Shahroudy, A., Shuai, B., Liu, T., Wang, X., Wang, G., Cai, J., Chen, T.: Recent advances in convolutional neural networks. J. Pattern Recognit. **77**, 354–377 (2018)
23. Cheng, J., Wang, P., Li, G., Hu, Q., Lu, H.: Recent advances in efficient computation of deep convolutional neural networks. Front. Inf. Technol. Electron. Eng. **19**(1), 64–77 (2018)
24. Habibi, A.H., Jahani, H.E.: Convolutional neural networks. In: Guide to Convolutional Neural Networks, pp. 85–130. Springer, Cham (2017)
25. Yamashita, R., Nishio, M., Do, R.K.G., Togashi, K.: Convolutional neural networks: an overview and application in radiology. Insights Into Imaging 1–19 (2018)
26. Pan, S.J., Yang, Q.: A survey on transfer learning. IEEE Trans. Knowl. Data Eng. **22**(10), 1345–1359 (2010)
27. Weiss, K., Khoshgoftaar, T.M., Wang, D.J.: A survey on transfer learning. J. Big Data **3**(9), 1–40 (2016)
28. Szegedy, C., Vanhoucke, V., Ioffe, S., Shlens, J., Wojna, Z.: Rethinking the inception architecture for computer vision. arXiv preprint: arXiv:1512.00567 (2015)
29. He, K., Zhang, X., Ren, S., Sun, J.: Deep residual learning for image recognition. arXiv preprint: arXiv:1512.03385 (2015)

# Information and Communication Technologies

# A Smart City's Model Secured by Blockchain

António Brandão[1]([⊠]), Henrique São Mamede[2]([⊠]),
and Ramiro Gonçalves[3]([⊠])

[1] UAb e UTAD, Lisbon, Portugal
ajmbrandao@gmail.com
[2] UAb - Universidade Aberta, INESC-TEC, Lisbon, Portugal
hsmamede@gmail.com
[3] UTAD - Univ. Trás-os-montes e Alto Douro, INESC-TEC, Vila Real, Portugal
ramiro@utad.pt

**Abstract.** The topic of Smart Cities involves various aspects of convergence and support to necessarily safe activities, which are dependent on reliable data and reliable sources. The answer could be provided by means of the adoption model of Blockchain technologies for this purpose, applied to the different characteristics of the city from the electronic governance, the contracting of products and services, and the sensing or data collection. The multiple IoT objects and the multiple networks, together with the Blockchain technology, may result in safer and more efficient cities. This paper analyzes and deepens the concept of Smart Cities using Blockchain technology as a platform for safety and reliability of the data. It is intended to reflect critically on this topic, looking through the different aspects involving this concept and the way in which it can implement. It analyzes the Smart Cities and its ecosystems with the generic data models, integrated, with quality data and trusted information.

**Keywords:** Smart Cities · Blockchain · Applications · IoT

## 1  Introduction

This paper analyzes and deepens the concept of Smart Cities, using Blockchain technology as a platform for safety and reliability of the data. In the perspective of this text, the Smart Cities topic include the concepts of digital cities, cyber cities, digital or virtual communities and cities of knowledge, associated mainly to the human kind, his privacy, governance, communities, their relationships and behaviors.

Our aim is to solve data integration issues, with the guarantee of integrity and reliability of the data, under a Smart City model supported by Blockchain as a way to control information flows and ensure the reliability of the data, information, and transactions.

The key concepts of Smart City in the perspective that the city is not only physical but also virtual and digital, presents new dimensions that interfere with various actors, such as government and public administration, institutions, companies, and citizens. The new policies of ecological, social and economic sustainability, mobility, transport, and innovation, imply greater civic participation, a greater intervention of the different economic and social agents and promote the intensive use of Information and

© Springer Nature Switzerland AG 2019
J. Mejia et al. (Eds.): CIMPS 2018, AISC 865, pp. 249–260, 2019.
https://doi.org/10.1007/978-3-030-01171-0_23

Communication Technologies (ICT) and new applications and Information System (IS) that should converge to common strategies.

These strategies must be based on models of integration, relationship, interaction, participation, access to information, involving robust connectivity infrastructures, fiber optic networks, wireless networks, and network integration that can support the growing needs of the various services and applications, with monitoring and control, based in sensors and IoT devices.

# 2 Key Concepts

A Smart City adds new concepts and diverse perspectives, with more data and information, procedures and processes, indicators and algorithms.

## 2.1 Smart City Concept

The concept of Smart City leads to unique and complex urban entity with three foundational bases - technological, ecological and social - oriented to an innovative, ecological city, cybernetics, "smart" information, and knowledge. The concept of a Smart City also contextualizes the technology to be used in systems and services for people [1].

The connection to people, social cohesion, heritage, health, lifestyle "smart", through knowledge, learning, innovation, creativity, and science, reveal a city of the future, in a sustainable world, as:

- The informed and "smart" communities, governments open to participation to new policy work, education, social, health and public safety-oriented citizenship;
- Integrated and efficient infrastructure, "smart" buildings, optimized communications, supply chains, sustainable public transport with enhanced mobility, and green energy;
- Retention and savings of water and the treatment and disposal of waste;
- Economic activity with the services, banks, tourism, trade, and industry, on-site promotion processes overall.

## 2.2 Smart Urbanism

The Smart Urban planning, integrating the concept of Smart City, is still not present in the development of the Smart Cities.

The new urban solutions involve the optimization and management of energy resources through the Internet of Energy (IOE) or the network of "smart" technology. This allows real-time monitoring and optimization of energy flows; security including CCTV systems, surveillance cameras, emergency response services and improved automated messages to alert citizens.

The real-time information on the status of the city should be available and present in the environment and transport that allow controlled levels of pollution, "smart" streetlights, congestion rules and new public transport solutions to reduce car use.

The zero energy buildings (ZEBs) is part of one of the pillars of Smart City, ecological and sustainable. The purpose of ZEBs should provide for active interaction with the urban power grids, using automated systems and parameter design, learning algorithms and sensitive sensor networks [2]. The energy consumption of the buildings use phase and energy efficiency criteria, the adoption of innovative approaches that promote a holistic approach, incorporating a combination of technologies and energy management solutions in real time, with social, life and economic viability [3].

The basic concepts and architectural components of Smart Cities pass-through power management (complying with ISO 50001), "smart" homes, vehicle networks, smart grids and quality of life (ISO 37120-Sustainable Community Development) [4].

**Building Information Modeling (BIM).** The BIM is the centerpiece of the new Smart City planning strategy. The BIM provide accurate and up-to-date building data and is an integrated approach to building lifecycle data management, with the Information to control the project, building, and maintenance [5].

The benefits are the projected city "equal" to the real city, with control over the compatibility aspects between specialties and interfaces with the information necessary for the integration and interconnection between different architectures. A city conceived, planned and sustained ecologically and economically studied to control the life cycle of its infrastructure and its equipment, operation and maintenance, and that applies and verifies the application of the rules, laws and best practice. Complementing the BIM concept by applying it to a Smart City, City Information Modeling (CIM) becomes an important resource to help improve quality of life in cities and for its structured planning [6].

## 2.3 Reference Architecture

The creation of a reference architecture for Smart City projects can serve as design language to create plans for Smart Cities, with a plan that meets different stakeholders, devices, platforms and technologies [7].

In order to design large and heterogeneous ecosystems such as the Smart Cities, it is necessary to be supported in a reference architecture, such as Service Oriented Architecture (SOA). It will serve as a plan and starting point that contains architectural blocks, through best practices and standards, with an initial meta-model with multiple views and especially in relationships to the Smart Cities software community [7].

**Possible Core Solution Architecture to Urban Planning.** Urban planning should provide a complete system consisting of various types of sensor deployment in multiple locations and ecosystems.

A four-tier architecture, as represented in Fig. 1, is proposed in [8], which includes:

- The level 0, IoT sources responsible for generation and collection of data;
- The level 1, which is responsible for all kinds of communication between IoT devices;
- The level 2, responsible for managing and processing data using a framework with Hadoop Spark;

- The level 3, in charge of the application and use of data analysis and results generated.

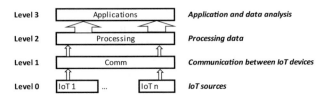

**Fig. 1.** Four-tier architecture

The implementation of this system consists of several steps that begin with the generation of data gathering, aggregating, filtering, sorting, pre-processing, computing and decision-making.

The future development of the city, historical data are analyzed offline with Hadoop, using the MapReduce programming. This system can be more scalable and efficient than other systems. Moreover, it presents a high efficiency measured in terms of production and processing time [8].

### 2.4  Natural Ecosystems in Smart Cities

The ecosystems of Smart Cities, driven by data, the IoT and the recent advances in data processing infrastructure, are providing a boost in its development, with the collection and analysis of reliable data for the "intelligence" urban, which depends on the ability to translate IoT data into useful and innovative services [8, 9]. As such, it is necessary to develop new design principles for the development of applications and must be from a simplified conceptual framework.

This framework describes the cities smart ecosystems, based on concepts of natural ecosystems, in a model whose objective is to present new "metaphors" to design IoT applications based on the observation of patterns of communication and interaction in natural ecosystems.

The availability of resources and the notion of data cycles must correspond to the nature of the resource cycles to ecosystems data-driven in cities can be directly visible and implementation with the application of appropriate regulations based in indicators [10].

The indicators, to measure the "behavior" of the cities, with the complexity of sustainable development and of the policies, are tools to evaluate and improve the cities. The choice of indicators and their practical use have the following challenges: the low availability of standardized data, open and comparable; and the lack of robust data collection institutions at the city scale to support monitoring. The technological changes, data management, and Smart Cities have increased the complexity as they add technologies, local initiatives and involve other natural ecosystems and institutions for data collection [11].

The vagueness of the objectives, the data constraints leave performance indicators subject to interpretations on the appropriate local implementation. The goals should be universal and their adaptation and interpretation should make them useful in various levels of the city, region, country, or supranational space through planning and targeted policies for citizens' well-being and the environment that it involves [11].

## 2.5    Sustainable Cities versus Smart Cities

The city's assessment tools are the support for decision making for urban development, responsible for providing assessment methodologies to cities, to show the progressive evolution toward defined goals [8].

We may change the term of Smart Cities to more accurate, Sustainable Smart Cities. Where the Smart Cities are in addition sustainable urban landmarks, through the evaluation of the Smart City performance using not only output indicators measuring the efficiency of the deployment of smart solutions, but also impact indicators to assess the contribution to the ultimate goals, such as environmental, economic and social sustainability [12].

The large number and the spread of Smart Cities settings pose challenges for setting goals for the cities and the concept of sustainability [12].

The role of technology in Smart Cities should be the sustainable development of cities, not the new technology as an end in itself and that a city that is not sustainable is not "smart". The sustainability assessment should be part of the development of Smart Cities, which will help ensure that sustainability is not neglected in the development of Smart City, environmental, economic or social [12].

The vision of the Smart City includes the complex coordinated activities; incentives such as flexible control mechanisms; and procurement and governance based on utility infrastructure of the Smart City. This will require the completion of a comprehensive Smart City platform and with the latest technologies that can be used to support the platform [13].

## 2.6    Blockchain - Features

Although the Blockchain technology can still be considered in its initial phase, its application will be in many sectors and will make processes more transparent, secure, and efficient.

The Blockchain technology is based on a distributed system to record and cryptographically store a consistent and unchanging linear record of transactions between network users [14]. The Blockchain, in its conception, is a public and distributed ledger that provides information to participants of the digital transactions that have already been executed. The major Blockchain features are shown the following.

In each block, there is a private key and a public key. This public key is shared with all others. When a new transaction is initiated, the future owner sends his or her public key to the original owner. The creation of new block strings is based on hash functions, where the output is the result of a transformation of the original input information. The hash function is an algorithm that receives an input and transforms it into an output, through a "cryptographic hash function", which is difficult to reverse or re-create the data entry from its hash value.

The Blockchain can be understood as a distributed "database", structured in a network, its result being visible by the nodes and according to the actions taken by the "users" of the network.

The Blockchain also eliminates the need for a reliable counterparty to guarantee or hedge a transaction. Transactions, though traceable, are available with the feature not to reveal your identity.

The main characteristics of Blockchain platforms go through the protocol of sending, receiving and registering value, through a collaborative effort, open source licenses and governance mechanisms and the inviolability of the system.

Thus, Blockchain technology can replace systems that depend on third parties, eliminates intermediaries and confirms third-party transactions, potentialized the process of disintermediation and with consensus mechanisms, is distributed to transmit value in a transparent and secure way.

A Smart City can use these Blockchain's characteristics and mechanisms. The consequence will be the "elimination" of intermediation, the connection directly between citizens and the government with the citizens, a new city, with more trust and more efficient.

# 3   Problems

In this section, we review the problems that we consider to be the most important and that goes through data processing, re-use of Smart Cities data, adoption of distributed cloud computing, Big Data, and IoT, and Security in Smart Cities.

## 3.1   Data Processing

The data processing for Smart Cities can take several steps from collecting data from different heterogeneous sources, real-time processing and delivery to services or applications used for optimization of transport systems, people management, energy management, management of water resources and air pollution management, which require different data processing techniques.

The text of [15] relates seven steps for processing data: the collection of data from different sources; data normalization; the tasks of data (treatment); data storage; data analysis; data visualization and decision support systems. The generic architecture for data flow management is described in functions and components for each stage, with specific technologies and auto-adaptive optimization methods used in architecture.

The Data must be processed with a certain frequency or in real time using big data techniques to extract the information needed for decision-making on the evolution of the city. The problem is the quality of the information used to inform citizens to take certain decisions and actions or activate actuators that allow automatic processes.

## 3.2  Reusing Data from Smart Cities

The platforms or portals of Smart Cities offer many data that can be used by public and private entities to create new services. This is the valuable source for the reuse of data. That will allow the treatment to various levels, business, academic, private and public, for deploying solutions or assessments that may create value for citizens and society.

The three-stage model, presented in [16], defines the following stages: Smart City data entry modeling, which includes several dimensions that make them attractive for reuse data; analyzes the mechanisms for creating innovative products and services; explains how these products and services affect your society.

Their modeling will allow cities and their government to promote services that create value, and released by the Smart Cities to society, which allows the creation of an ecosystem of agents that reuse. In this way, we can improve the quality of life of citizens by creating and optimizing these ecosystems and the economic and social agents, re-using the data to create innovative services. The participation of citizens acted as "sensors" urban, close to reality, improve the real expectations of the citizens, optimizing investments or predicting the potential impact on citizens to redesign the services, with quality, security, privacy, and ethics [16]. The problem is the quality of the data used and reused.

## 3.3  The Distributed Cloud Computing

A Smart City architecture organized in layers includes a layer of detection that generates the data, a network layer to move the data, a middleware layer that manages the data collected and prepare them for use and application layer that provides "smart" services that enhance data [2]. The amount of information to be generated in a Smart City is high and growing, adding data from other sources, such as mobile devices, surveillance cameras or web services. It is necessary to study more distributed solutions with the nearest data repository of real-time applications. The problem is how an implementation of cloud computing technologies can partially process the data generated close to the source sensor and transfer only the relevant information, to promote a distributed computing environment in the city used in order efficient, scalable and dynamic [2].

## 3.4  Big Data and IoT

The expansion of Big Data and Internet technologies (mainly IoT) play an important role in the feasibility of Smart Cities initiatives. The Big Data has the potential to achieve for the Smart Cities of the relevant information, through the treatment of a large amount of data collected from various sources. The IoT allows the integration of sensors in real and network integrated environment. The combination of IoT and Big Data allow solve problems related businesses and technology to update the vision, principles and application requirements of Smart Cities. How the context of data management, with the technologies used in Smart Cities, with business models and the underlying architecture can play an important role in relevant information and value for decision-making [17].

The Big Data will allow analysis of city "life", through IoT's data, consolidated in generic data models, to define new models of urban governance and new options for cities more sustainable, competitive, productive, efficient, transparent and open [18].

### 3.5 Security in Smart Cities

The proliferation of technologies from the Internet of Things (IoT), cloud computing and the interconnection of networks, enable Smart Cities to foster innovative solutions with a more direct interaction with citizens. This ease and functionality also bring challenges of security and privacy of information, focusing on the need to provide a security framework between the systems that process and systems that collect, and so provide the devices with more "intelligence" for a more secure communication. The problem is how a guarantee the security on all levels.

**Other Protection Mechanisms of Smart Cities.** A Smart City needs scalable and lightweight authentication structures that protect from internal and external intruders. They can be followed by several standards, such as the case of the IEEE 1609.2 v2 (Standard for Wireless Access in Vehicular) that specifies a set of security services that support e.g. vehicular communications.

The solutions based on public key infrastructure (PKI) can check the authenticity, but cannot lead to a scalable solution. It is necessary to investigate innovative solutions for messaging that can ensure the privacy and anonymity of new strategies for a "smart" urban environment.

The desired security properties undergo a light authentication, scalability, security architecture and resilience against malicious attacks us and Deny of Service (DoS).

## 4    Solution

The risk system is the commitment of the information, given that everything is interconnected, from the water supply system, energy system, public transport and others critical infrastructures. The big advantage of integration and interconnection is also its great vulnerability and that can affect the security of citizens [4].

The adoption of technologies such as Blockchain can come to solve this difficult problem. The integration of Blockchain technology with distributed devices in a Smart City will create a single platform, with the devices to communicate securely in a distributed environment, creating a model that allows integrating different systems and platforms through interoperability and scalable mechanisms [19].

This perspective is reinforced growing importance of new research areas, in addition to the crypto-coins, referred to in the literature review [20], where articles on IoT are verified, with 28% of applications in the financial area, with 14%, e-governance with 12%, smart contracts with 10%, Smart Cities and businesses with 9% and health with 5%. These new areas clearly apply to the operation of a Smart City.

### 4.1    The Model

We propose a Smart City's Context Model, as represented in Fig. 2 that intends to organize the city in ecosystems, ecosystems in application domains, application domains in generic data models and data models in IoT devices.

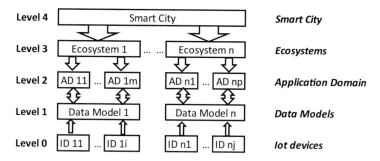

**Fig. 2.**  A Smart City's Context Model

This model has 5 levels that correspond to 5 stratifications of aggregation and summary, providing the upper level with information necessary and sufficient, and to guarantee the management of the downer level.

The problems will be treated in each level they are to solve the needs of data processing, with the distributed cloud computing, the Big Data, the reusing data from Smart Cities, the IoT Data, and the Security in Smart Cities.

The Blockchain applications are typified in six groups, as described in Table 1, which intend to characterize the ways in which they should interact with the model.

**Table 1.**  The Blockchain's application types

| Type | Applications |
|------|--------------|
| BC1 | *Secure Transactions* |
| BC2 | *Data Security* |
| BC3 | *Data Flow Control* |
| BC4 | *Accepted Devices* |
| BC5 | *Version Control* |
| BC6 | *Systems security* |

In Table 2, it is intended to review for each level of the model the applications that should be used.

The proposal intends to apply Blockchain base applications applied according to Table 2, at each level, that address issues of security, reliability, data processing, handling of Big Data and IoT Data, and the secure connection to distributed cloud computing.

**Table 2.** Blockchain's application in Smart Cities context

| Smart City context | | Blockchain application | | | | | |
|---|---|---|---|---|---|---|---|
| | | BC1 | BC2 | BC3 | BC4 | BC5 | BC6 |
| Level 0 | Smart City | x | | | | | |
| Level 1 | Ecosystems | x | | | | | x |
| Level 2 | Application Domain | x | x | x | | x | x |
| Level 3 | Data Models | | x | x | x | x | x |
| Level 4 | IoT devices | | | x | x | x | x |

**Other Solution.** As described in [21], a possible solution can be supported by the system, as the main components are the Blockchain, Smart Contracts, Intermediate Servers and an Off-chain Database. The main parties of the system are Publishers and Subscriber and the main end parties are IoT devices and Users. In this implementation, the system can be created a private Blockchain, with a couple of nodes, and notification smart contracts on new data updates.

## 5    Conclusions

The city is not only physical but digital or virtual where critical factors such as governance, innovative policies, ICT, ecological, social and economic, civic participation, the involvement of economic actors, mobility policies and transport converge on common strategies and approaches emerging as BIM and CIM. But it's not enough to understand and plan the city; it's necessary to promote the information's quality, to ensure trustfully data, with security and responsibility.

The information's quality, as result of trustful data, can use Blockchain platforms, in comprehensive and critical data models, becoming a viable and reliable option, to promote trust data and security.

Natural ecosystems translate into ecosystem data that build the technology more efficient ways to manage and controlled the cities of the future, with needs of security and the trust, perhaps based in Blockchain. Many research will be developed to be a reality.

The model proposed in this article is oriented to organize an intelligent city in ecosystems, supported in several types of Blockchain applications, in order to try to solve the problems identified.

As future work, we intend to research of solutions and applications that use Blockchain as a base technology, to guarantee the immutability of the records, the redundancy of the data and the confidence in the information supported in reliable data.

The validation of the model proposed will be done in the research to develop in a specific ecosystem, in an application domain, with a certain generic data model, loaded by IoT device data.

The research, in science and technology Web, has a new challenge, how to become the IoT architecture, supported in Blockchain, more efficient to allow improve the collection of reliable urban data. The new smart objects will be connected to cloud computing and converge data to platforms aggregated in Information, with news inputs for the algorithms and Artificial Intelligence (AI) systems, to form the decisions.

# References

1. Jong, M., Joss, S., Schraven, D., Zhan, C., Weijnen, M.: Sustainable–smart–resilient–low carbon–eco–knowledge cities; making sense of a multitude of concepts promoting sustainable urbanization. J. Clean. Prod. **109**, 25–38 (2015)
2. Kylili, A., Fokaides, P.A.: European smart cities: the role of zero energy buildings. Sustain. Cities Soc. **15**, 86–95 (2015)
3. Lee, J.S., Kwon, O., Choi, S.: Identifying multiuser activity with overlapping acoustic data for mobile decision making in smart home environments. Expert Syst. Appl. **81**, 299–308 (2017)
4. Khatoun, R., Zeadally, S.: Smart cities: concepts, architectures, research opportunities. Commun ACM. **59**, 46–57 (2016)
5. Wolisz, H., Böse, L., Harb, H., Streblow, R., Müller, D.: City district Information modeling as a foundation for simulation and evaluation of smart city approaches. 9
6. de Amorim, A.L.: Cidades Inteligentes e City Information Modeling. In: XX Congreso de la Sociedad Iberoamericana de Gráfica Digital (SIGraDi), pp. 481–488. Editora Blucher, Buenos Aires (2016)
7. Abu-Matar, M.: Towards a software defined reference architecture for smart city ecosystems. Presented at the 2016 IEEE International Smart Cities Conference (ISC2), Smart Cities Conference (ISC2), 2016 IEEE International, 1 September 2016
8. Rathore, M.M., Ahmad, A., Paul, A., Rho, S.: Urban planning and building smart cities based on the Internet of Things using big data analytics. Comput. Netw. **101**, 63–80 (2016)
9. Schaffers, H., Komninos, N., Pallot, M., Trousse, B., Nilsson, M., Oliveira, A.: Smart Cities and the Future Internet: Towards Cooperation Frameworks for Open Innovation. In: Domingue, J., Galis, A., Gavras, A., Zahariadis, T., Lambert, D., Cleary, F., Daras, P., Krco, S., Müller, H., Li, M.-S., Schaffers, H., Lotz, V., Alvarez, F., Stiller, B., Karnouskos, S., Avessta, S., Nilsson, M. (eds.) the Future Internet, pp. 431–446. Springer, Berlin (2011)
10. Sinaeepourfard, A., Garcia, J., Masip-Bruin, X., Marin-Tordera, E., Cirera, J., Grau, G., Casaus, F.: Estimating smart city sensors data generation. Presented at the 2016 Mediterranean Ad Hoc Networking Workshop (Med-Hoc-Net), Ad Hoc Networking Workshop (Med-Hoc-Net), 2016 Mediterranean, 1 June 2016
11. Klopp, J.M., Petretta, D.L.: The urban sustainable development goal: indicators, complexity and the politics of measuring cities. Cities **63**, 92–97 (2017)
12. Ahvenniemi, H., Huovila, A., Pinto-Seppä, I., Airaksinen, M.: What are the differences between sustainable and smart cities? Cities **60**, 234–245 (2017)
13. Dhungana, D., Engelbrecht, G., Parreira, J.X., Valerio, D., Schuster, A., Tobler, R.: Data-driven ecosystems in smart cities: a living example from Seestadt Aspern. Presented at the 2016 IEEE 3rd World Forum on Internet of Things, WF-IoT 2016, 6 February 2017
14. Risius, M., Spohrer, K.: A Blockchain research framework: what we (don't) know, where we go from here, and how we will get there. Bus. Inf. Syst. Eng. **59**, 385–409 (2017)
15. Dustdar, S., Nastic, S., Scekic, O.: A novel vision of cyber-human smart city. Presented at the 2016 fourth IEEE Workshop on Hot Topics in Web Systems & Technologies (HotWeb), 1 January 2016

16. Chilipirea, C., Petre, A.-C., Groza, L.-M., Dobre, C., Pop, F.: An integrated architecture for future studies in data processing for smart cities. Microprocess. Microsyst. **52**, 335–342 (2017)
17. Hashem, I.A.T., Chang, V., Anuar, N.B., Adewole, K., Yaqoob, I., Gani, A., Ahmed, E., Chiroma, H.: The role of big data in smart city. Int. J. Inf. Manag. **36**, 748–758 (2016)
18. Kitchin, R.: The real-time city? Big data and smart urbanism. GeoJournal **79**, 1–14 (2014)
19. Biswas, K., Muthukkumarasamy, V.: Securing smart cities using Blockchain technology. Presented at the 2016 IEEE 18th International Conference on High Performance Computing and Communications; IEEE 14th International Conference on Smart City; IEEE 2nd International Conference on Data Science and Systems (HPCC/SmartCity/DSS), High Performance Computing and Communications; IEEE 14th International Conference on Smart City; IEEE 2nd International Conference on Data Science and Systems (HPCC/SmartCity/DSS), 2016 IEEE 18th International Conference on, HPCC-SMARTCITY-DSS, 1 December 2016
20. Brandão, A., Mamede, H.S., Gonçalves, R.: Systematic review of the literature, research on Blockchain technology as support to the trust model proposed applied to smart places. In: Rocha, Á., Adeli, H., Reis, L.P., Costanzo, S. (eds.) Trends and Advances in Information Systems and Technologies, pp. 1163–1174. Springer, Cham (2018)
21. Rifi, N., Rachkidi, E., Agoulmine, N., Taher, N.C.: Towards using blockchain technology for IoT data access protection. In: 2017 IEEE 17th International Conference on Ubiquitous Wireless Broadband (ICUWB), pp. 1–5. IEEE, Salamanca (2017)

# Framework for the Analysis of Smart Cities Models

Elsa Estrada[1](✉), Rocio Maciel[2], Adriana Peña Pérez Negrón[1],
Graciela Lara López[1], Víctor Larios[2], and Alberto Ochoa[3]

[1] CUCEI of the Universidad de Guadalajara,
Av. Revolución 1500, Col. Olímpica, 44430 Guadalajara, Jalisco, Mexico
{elsa.estrada,graciela.lara}@academicos.udg.mx,
adriana.pena@cucei.udg.mx
[2] CUCEA of the Universidad de Guadalajara,
Periférico Norte 799, 45100 Zapopan, Jalisco, Mexico
{r.maciel.mx,victor.m.lariosrosillo}@ieee.org
[3] Universidad Autónoma de Ciuadad Juárez,
Av. Plutarco Elías Calles no. 1210, 32310 Ciudad Juárez, Chihuahua, Mexico
alberto.ochoa@uacj.mx

**Abstract.** Smart cities evolution forces auto adjustments. A constant change that difficult methodologies and tools development aimed to measure and evaluate the huge number of variables involved. The Smart City metrics model is composed by its determined key performed indicators (KPI); with different aims a number of models have been proposed by different organizations, which difficult its comparison. In this paper, we propose a framework to apply Data Science to KPIs from Open Data. This framework is organized by a set of tools: a KPI tree structure; a JSON document; a web app with non-supervised or supervised knowledge for the models evaluation; and the infrastructure for reports reception and attention. In such a way that this framework creates an infrastructure that goes from the treatment of Open Data to models evaluation and its management.

**Keywords:** Data Science · JSON · Open Data · KPI · Smart City

## 1 Introduction

According to [1] a Smart City is an emerging strategy to mitigate the growing urban population and its fast urbanization consequences, and it has to be focused on its key performance indicators (KPI) characterization. Other angle for a Smart City is the one presented by Towsend [2], focused on operations and benefits. He refers to a Smart City that solves problem events through monitoring by using sensors [3].

The Smart City concept has also being related to Sustainable City [4], approach that emphasizes the Information and Communication Technologies (ICT) as a dominant

---

The original version of this chapter was revised: The author name "Graciela López Lara" has been changed to "Graciela Lara López". The correction to this chapter is available at https://doi.org/10.1007/978-3-030-01171-0_27

J. Mejia et al. (Eds.): CIMPS 2018, AISC 865, pp. 261–269, 2019.
https://doi.org/10.1007/978-3-030-01171-0_24

element to improve urban factors such as life quality, urban operation and services, and its economical, social and environmental supply.

Therefore, it can be said that the Smart Cities are a strategic resource to solve the problems that affect their sustainability and the quality of life of the population, promoting the development of instruments to monitor and to evaluate the urban phenomena. A well known Smart Cities tool is the metric model, it helps to evaluate the cities, along with the autonomous systems that implement diverse types of components that include Internet of Things (IoT), Software Defined Network (SDN) or the 5G transfer standards, and Data Science algorithms and intelligent systems to support the decision-making.

## 1.1  Smart Cities KPI Models

Metrics tools for Smart Cities support decision-making and performance improvement in issues such as welfare and quality of life. These are abstract and intangible aspects, so they cannot be directly measured. Their measurement is expressed based on other indicators like the percentage of 15 years old population with primary school, or the percentage of emissions with greenhouse effect. According to Lang [5], metrics measurement for cities main goals are diagnosis and evaluation, conditions comparison, support and persuasion, Pedagogy, and generation of trust in life. Smart Cities are then interested in establishing the quality of life in order to identify areas for improvement as they support the decision-making.

Governments are paying attention to these instruments to understand the citizen's opinion about the factors that give them welfare. To this end, a number of models that give structure to the KPIs have been presented. Boyd Cohen, who formed an IBM's representatives committee, Smart City Expo, Smart Cities Council and Innovation Centre in Cities, generated a model to measure cities with six main KPIs, with 18 indicators subdivided in 62 [6]. Sixteen of them were mapped by ISO 37120 [7] which evolved to its model of 17 KPIs with 100 indicators. The Global Power City evaluates and classifies the best cities according to 6 KPI with 26 indicators subdivided in 70 [8]. Other models have appeared with the Human Development Index [9] or the ecologic path aimed to calculate the city consumption resources and the waste residual.

## 1.2  Smart Cities Data Sources Structures

The Smart City involves a huge volume of data, which demands a highly efficient treatment for its analysis. The factors related to this expected amount of data are the five Vs, these are volume, velocity, variety, value and veracity [10], commonly used to characterize different aspects of Big Data which have been increased to 10 due to the addition of other factors such as visualization, volatility, vulnerability, validity and variability [11].

One of the first tasks is to detect the target of evaluation, that is, functions, categories or areas in indicators to gauge from environment to the level of services requirements. This implies finding measures to understand the city performance, discover underlying trends, compare characteristics and identify strengths and weaknesses of the city in a comparative way, in order to measure its development.

Therefore, patterns from the city's KPI analysis required to be extracted. Scientific analysis is one of the tools to assist in this process, together with the algorithms to classify the information, such as the Bayesian model, especially the useful kernel for classification tasks, regression analysis, and cluster analysis of spatial data, among others. Performing the correlations between the KPIs leads to the establishment of forecasting via the observation of patterns.

There are a number of tools for processing large volumes of data. Their selection will be based on the design of the outline of the databases; the desired attributes and its escalation are required to be specified. The existing outlines are limited to the extractions of chains; this prevents the analysis of the associated factors to other data type. Because the relational database models are not large enough to handle the extensive variety of formats and volume, a NoSQL scheme was designed with storage in columns to facilitate the scalability and speed of searches (For further details see [12, 13]).

## 2   Framework for Open Data Analysis

Based on two previous studies [14, 15], it was established that the measures for a Smart Cities are focused on three elements: (1) the management and control of the KPIs to configure and organize the metrics model; (2) the pattern discovery and its diffusion toward non-supervised or supervised, and georeferenced data; and (3) an intelligent dispatcher for service request based on the configuration and the observed patterns.

For these three elements, a framework composed with the next implementations is proposed:

(A)   A tree structure based manager of KPIs;
(B)   A designed JSON document to automatize the services dispatch, which will guide the monitoring of threshold values;
(C)   Four Web scenarios for Smart People for the City evaluation:
    (1)   The scenario for charging and mapping the content,
    (2)   The scenario for Data Science application with the reached pattern,
    (3)   The scenario for the object's individual track with respect to its distance to the threshold, and
    (4)   The scenario to monitor and dispatch service petitions; and
(D)   The infrastructure to support reception and attention reports from the different government entities involved.

Figure 1 shows these framework implementations, which will be described in the next sections.

### 2.1   KPI Tree Structure

The shape of indicators in the Open Data neither has a standard on its types definition, nor in their ranges or in units for measurement. They have no structure in the key-value. From a review on the applied models to measure citie's sustainability, quality of life, human development or the ability to consume what is produced, it was observed that they are typically subdivided for better expressing measurement accuracy. A nested structure was identified, along with groups by levels denominated layers.

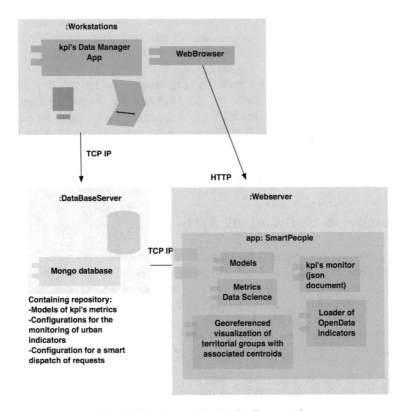

**Fig. 1.** Implementations in the Framework

A first implementation for handling these layers was introduced in [16]. In a new version of the application, the JSON documents are stored in a MongoDB. A next layer assigns a selected KPI as father, and so on. The list of the KPIs tree is returned in a chain format in the JSON document structure and each one is inserted in a cycle in the MongoDB for the connection with other methods.

The document conversion has two activities. The first one is to generate the JSON document for the root object. The second one is to generate a document for each KPI within the nested ones. Then its interface allows creating new models to be compared.

## 2.2   JSON Documents to Automatize Metrics for Service Dispatch

In large cities, the excess of applications for services is growing, the public and private entities as well as the government recurs to the monitoring of the indicators with the expectative to provide attention to all the citizen's petitions, such as the transportation demand, electric power, quality of the environment, hospitals, medications, definite constructions or quality of education among others. It is then desired to contribute with control of models of metrics to automate the detection and delivery of services throughout a JSON document (see Fig. 2). It functions as a monitor in which it relates each indicator or layer to a public service, connecting an event to the threshold values, the format of the monitor that allows the configuration according to the different indicators, as well as the type of method of analysis that is desired to apply.

```
{
"_id":"Monitor_PrimarySchools_OpenData",
"TerritorialEconomicDivisions_numberofcluster
                            s":"3",
"Algorithm":"Kmeans",
"Threshold":
       {
       "2016_E_Spanish:"",
       "2016_E_Math_Exc" :"",
       "2015_L_Math":"",
       "2015_M_Math":"",
       "2015_E_Math":"",
       "computers_working":"",
       "computers_with_internet":"",
       "computers_working_ineducation":"",
       "total_graduades":""
       },
  "DispatchingOfServicesConfiguration":
       {
       "2016_Esp_Exc":"SpanishAdvisors",
       "2016_M_Exc" :"Math Advisors",
       "2015_L_Mat":"MatAdvisors",
       "2015_M_Mat":"MathAdvisors",
        "2015_E_Mat":"MathAdvisors",
       "computers_working":"FinanzasSEP",
       "computers_with_internet":"FinanzasSEP",

"computers_working_ineducation":"FinanzasSEP",
       "total_graduades":
          {
          "Departmen_1": "ScholarshipUnit",
          "Department_2": "DIF"
          }
       },

"Territorial_division_1":
{
"Quality_of_Primary_Schools":
   {
   "14EPR0237K" :
      {
      "Description":
         {
         "Name":"",
         "Latitude":"",
         "Longitude":""
         },
      "2016_E_Spanish:"",
      "2016_E_Math_Exc" :"",
      "2015_L_Math":"",
      "2015_M_Math":"",
      "2015_E_Math":"",
      "computers_working":"",
      "computers_with_internet":"",
      "computers_working_ineducation":"",
      "total_graduades":""
      },
   "14EPR0293C" :
      {
      "Descrition":
         {
         "Name":"",
         "Latitude":"",
         "Longitude":""
         },
      "2016_E_Spanish:"",
      "2016_E_Math_Exc" :"",
      "2015_L_Math":"",
      "2015_M_Math":"",
      "2015_E_Math":"",
      "computers_working":"",
      "computers_with_internet":"",
      "computers_working_ineducation":"",
      "total_graduades":""
      }
   }
"total_population_age_primary_completion":"",
"PorcentageOfStudentswithPrimaryShcools": ""
}
```

**Fig. 2.** JSON Document: city's smart monitor

In the example of the JSON document the monitor of schools is implemented, where the 3 value corresponds to three clusters, equivalent to 3 economical territorial divisions resulting of tests of unsupervised learning applied to the Open Data to Guadalajara primary schools with unstructured format. The algorithm defined is kmeans because the values in this case are not labeled. The object nested Threshold has a structure to store the top values of each variable to monitor, DispatchingOfServicesConfiguration specifies the government entity that is responsible of the delivering of resources and services. Territorial_division_1 is a cluster example, Quality_of_Primary_Schools is the layer to evaluate by cluster, and 14EPR0237K is an id of a school that is a grouped scale in that cluster.

JSON objects establish the form of storage in MongoDB. A document corresponds to each unity to be measured; it can be a sensor or a school for example. The structure allows to storage each unity indicators. It also describes the mapping of unities of services related in such a way that the "Analysis of metrics" component uses the JSON document to visualize the dispatch orders queue (see Fig. 3) according to the threshold configured value for each indicator.

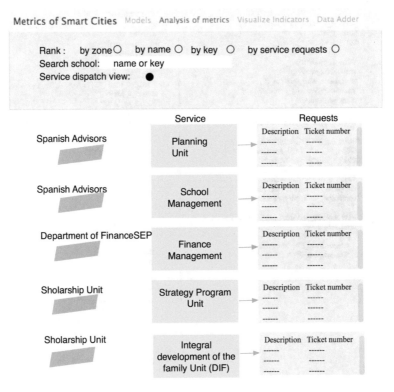

**Fig. 3.** Implementation of monitor and dispatch petitions. Case of study quality of education in primary schools inside Open Data

### 2.3 Evaluation Models Algorithms: Supervised, Unsupervised Learning and Means

A tool programmed in Java, JavaScript, and Webservice, with connection to MongoDB to manage the collections in JSON format was developed. This Smart Visualization tool permits the observation of territorial patterns to simulate the citizen abilities, to make inferences in city KPI. This simplifies the activity analysis through Open Data from smart tracking and citizens' tools, identifying marginalized areas generating awareness of the city deficiencies.

A georeferenced Open Data through Machine Learning is performed. In the tool the selected KPIs are storage, with the Model name as the root of the tree. Then an exploration of the Model is performed, for that, a preliminary Data Mining that joins tables and files also performs a filtering and cleaning of the data. Afterwards, a territorial dependency for the trends' observation is performed creating autonomous learning cluster techniques for different number of zones and variables. Then the software components are designed, programed and integrated. A case of study for Education Sustainability following this process can be found in [14]. This case manages non-supervised learning, k-means specifically see Fig. 4; this Figure shows k-means applied to KPI in primary school of Guadalajara city. Three figures grouped by color can be observed in this map: green, blue and yellow as a result of the k-mean approach for nine sectors (clusters) of the city with an 80% sensibility. The three types of centroid schools are placed in the top of the map. Also a case of study for air pollution with supervised learning can be found in [17].

## Schools of de Guadalajara clustered by zone

| Indicators | Type 1 | Type 2 | Type 3 |
|---|---|---|---|
| '2013_Exc_Esp_6' | 5.69% | 8.01% | 3.29% |
| '2013_Mat_Exc_6' | 14.41% | 18.3% | 7.46% |
| '2013_Exc_Esp_5' | 6.41% | 6.33% | 2.34% |
| '2013_Mat_Exc_5' | 17.01% | 18.25% | 7.81% |
| '2013_FyC_Exc_5' | 3.05% | 3.5% | 1.23% |
| '2013_Exc_Esp_4' | 8.4% | 9.52% | 2.5% |

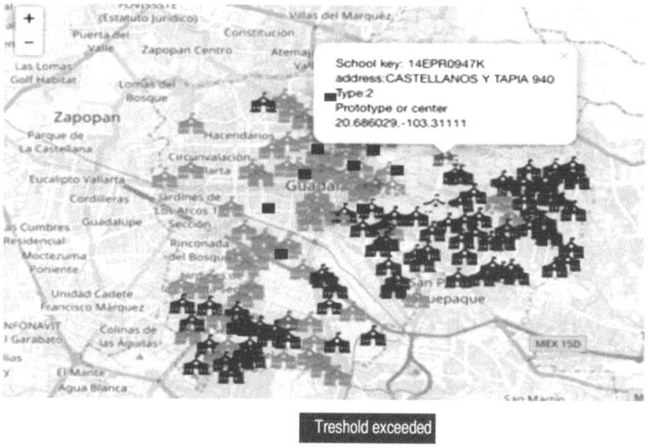

Treshold exceeded

**Fig. 4.** Georeferenced data interface to identify threshold parameters in order to dispatch intelligent services.

This part of the tool contents the scenario for charging and mapping the contents; that is the scenario for Data Science application with the reached pattern, and the scenario for the object's individual track with respect to its distance to the threshold.

For supervised data such as data from the air pollution, a Java algorithm was developed to the creation, measurement and writing in real time of these documents. The algorithm includes the extraction of environmental records with CVS format files. It also converts these files to a list form and processes the information produced by each sensor, which allows handling the variety in the indicators that each sensor sends in the monitoring.

The algorithm considers that the same indicator does not send continuous measurements or information, thus the mean is calculated only with the data that appears in the CVS file. The calculation of the global means of the indicators is stored by hour, by the sensors. Since each of them may possess different indicators, the algorithm calculates the mean of each one that generates a measure, and with the last hour in which it had been updated (See [15] for a more detailed explanation).

### 2.4    Protection Mechanisms of Smart Cities

Security is an important aspect to include in the development of Smart City frameworks. Because the transactions to implement require the exchange of information from and to the net, as unloading and loading of files, transfer of services petition tickets, receiving and following up of report status, so that is essential the use of strategies such LDAP, PKI, SHA3, to guarantee the protection, confidentiality, files integrity and the activities over by the auto dispatching.

## 3    Conclusions

This framework works well for supervised and non-supervised learning. The intelligent aspect of these tools is in the pattern identification through Machine Learning. The interface for the georeferenced data visualization, helps to observe each zone and can adopt a different pattern creating shortage alerts that help to automatize certain demands in the region. The JSON documents are the baseline to configure the type of algorithm for the analysis, and the flow for the dispatch of services, and the threshold values. However, this assumes a servers' infrastructure of large-scale, organization changes in government departments, and a coordinated communication among server clusters.

As future work the tool will be implemented in the Public Education Secretariat in Guadalajara.

## References

1. Airaksinen, A., Pinto Seppä, I., Huovila, A., Neumann, H.-M., Iglar, B., Bosch, P.: Smart City performance measurement framework. Presented at the February 6 (2018)
2. Townsend, A.M.: Smart Cities (2013)
3. Ahlgren, B., Hidell, M., Ngai, E.C.H.: Internet of Things for smart cities: interoperability and open data. IEEE Internet Comput (6), 52–56 (2016)
4. Arcadis: Sustainable Cities index (2016). http://www.arcadis.com

5. Moonen, T., Clark, G.: The business of cities 2013. 1–224 (2013)
6. Cohen, B.: Boyd Cohen. https://www.smart-circle.org/smartcity/blog/boyd-cohen-the-smart-city-wheel/
7. Claude-Anne, W.: Sustainable development and resilience of communities—indicators for city services and quality of life. 1–92 (2013)
8. Institute for Urban Strategies the Mori Memorial Foundation: Global Power City 2017 (2017)
9. United Nations Development Programme: Human Development Report. 1–286 (2017). hdr.undp.org
10. Ishwarappa, Anuradha, J.: A brief introduction on Big Data 5 Vs characteristics and Hadoop technology. Procedia Comput. Sci. **48**, 319–324 (2015)
11. tdwi ed: UPSIDE where DATA means BUSINESS
12. Edlich, S.: 2011 The NoSQL year. http://www.google.com.mx
13. Brewer, E.: CAP twelve years later: how the "rules" have changed. Computer **45**(2), 23–29 (2012)
14. Estrada, E., Ochoa, A., Bernabe-Loranca, B., Oliva, D., Larios, V., Maciel, R.: Smart City visualization tool for the Open Data georeferenced analysis utilizing machine learning. Int. J. Comb. Optim. Probl. Inform. **9**(2), 25–40 (2018)
15. Estrada, E., Maciel, R., Ochoa, A.: Best practices to implement a NoSQL method for the Smart Cities metric analysis
16. Estrada, E., Maciel, R., Gomez, L.: NoSQL method for the metric analysis of Smart Cities (2015)
17. Estrada, E., Kalichaning-Balich, I., Martinez, P., Mora, O.B., Maciel, R.: A parallel support vector machine algorithm to identify patterns of pollution in Smart Cities for metropolitan zone of Guadalajara (2017)

# Performance Analysis of Monolithic and Micro Service Architectures – Containers Technology

Alexis Saransig[1] and Freddy Tapia[2(✉)]

[1] Universidad Técnica del Norte,
Av. 17 de Julio, 5-21, 100105 Ibarra, Ecuador
afsaransigc@utn.edu.ec
[2] Universidad de las Fuerzas Armadas ESPE,
Av. General Rumiñahui s/n, 171-5-231B, Sangolquí, Ecuador
fmtapia@espe.edu.ec

**Abstract.** Comparative analysis of the performance of hardware resources, between Monolithic Architecture and Micro services Architecture, using virtualization technology based on development and production environments. Today, the new trend is the development and/or deployment of applications in the Cloud, in this aspect, monolithic applications have flexibility, scalability, maintainability and performance limitations. On the other hand, the focus of Microservices adapts to new trends and solves these limitations. Meanwhile, virtualization with virtual machines is currently not efficient enough with hardware resources. With the appearance of containers, this problem is solved due to its functioning characteristics as independent processes and resources optimization. Now, two scenarios are presented, the first consisting of a Web application based on a Monolithic Architecture that is executed in a Kernel based Virtual Machine - KVM and the second scenario shows the same Web application, this time, based on a Micro services Architecture and running in containers. Each scenario is subjected to the same stress tests; the generated data are recorded in "log" files for further analysis. The hardware resources are the same for both scenarios. The comparison of these scenarios helps to identify the efficiency of the Application and the hardware resources, as well as the development and/or deployment of Applications. This can be improved with the use of Microservices and Containers. In addition, the reduction of costs that would imply the optimization in the resources. For greater reliability in the interpretation of the data, two analysis tools were used: JMeter and NewRelic. Finally, the two resulting cases from the analysis are shown, each case being considered due to the feasibility of the same depending on the needs and availability of resources.

**Keywords:** Monolithic Architecture · Micro services · Containers
Performance

## 1 Introduction

Technological evolution aims to be more efficient in the use of resources. Software production must considerer different types of architectures, so that each product, would fulfill its objectives and be efficient in the use of resources. Monolithic Architecture is

J. Mejía et al. (Eds.): CIMPS 2018, AISC 865, pp. 270–279, 2019.
https://doi.org/10.1007/978-3-030-01171-0_25

an example of this case and commonly used in Software production. Its fusion with Virtual Machines has turned it into a successful and effective formula for small and large-scale projects. In general, a monolithic application declines in its performance once it starts to grow more than expected. The solutions can be varied, for instance, migration to new technologies, management of independent services, and more powerful servers. In the long term, it can generate high costs due to their limited capacity of escalation and maintainability. Innovation has given rise to new architectures that propose optimal solutions to improve the Software production process [12].

Micro services Architecture arises to provide solutions and to be a deciding factor when an appropriate decision, as to DevOps concerns and future projects that relate, for benefits this presents. The Microservices Architecture replaces the Monolithic with a lightweight, narrowly focused distributed system and isolated service [13].

Container technology is still little known in our environment. However, trends and specialized reports show that there is more acceptance for this technology since it provides a more efficient resources management compared to Virtual Machines.

In this research, a comparative performance analysis is made between applications with Monolithic Architecture running on a Virtual Machine against the same application, but in a version based on a Micro services Architecture using Containers. Both scenarios will be running on a computer with the same characteristics, they were subjected to stress tests, in order to analyze the data stored in "logs" files.

With the obtained results, decision-making is facilitated in terms of the efficient management of resources and the production of Software. In addition, benefits related to new development methodologies, that are not detailed in depth in this investigation are set. However, they remain open for further investigation.

## 2 Background

### 2.1 Monolith Architecture

It usually uses a single technology development, which limits the availability of a suitable tool for each task to be executed by the system. A single logical executable [1], in which any change made in a part of this type of system involves the construction and deployment of a new version of the entire system. It involved mostly aspects or layers like presentation, processing and storage in a single software component that is supposed to run on a single server.

The advantages are few; one of them is the minimum requirement of changes that can be made in their context. Meanwhile, its disadvantages fall in different areas such as maintenance, debugging, scalability, distribution and deployment. The efficiency is quantified due to the minimal changes that are made in its context; meanwhile, its shortcomings lie in the difficulty of performing maintenance tasks, debugging, scaling, distribution and implementation.

Until now, the technological trend has been focused on the development of Software, offering solutions focused on a Monolithic Architecture. Habit or facility based on knowledge, leads to trust the same old solutions even knowing the short or long term risks that can be found.

Every computer system or technological implementation is bound to evolve, just as solid disks have plunged punch cards. The level of complexity of modern systems requires improvements in both the production and the performance of the Software. This means that in the Monolithic Architecture have been found inevitable defects [2], which over time will play against it. This leading to modern architectures, such as Micro services.

### 2.2 Microservices Architecture

It is a relatively new architecture without formal definition yet. It is used to build large, complex and scalable applications. It is composed by small, independent and highly recoupable processes, communicating each one through API's [3].

With the features mentioned above, not only a solution is given to almost all the weak points of a Monolithic Architecture, but also it even proposes efficient solutions in the short and long term.

Functions composed of small and individual services, running in their own process and communicating with light application mechanisms. The independence of each one becomes fault tolerant and increases their availability. It is a new culture of automation, with decentralized processes, that allow independent deployments. Therefore, the whole is modeled around the core of the business [4–6].

Architecturally the structure of an application based on Micro services, differs in a great level from the Monolithic Architecture (Fig. 1 shows an example).

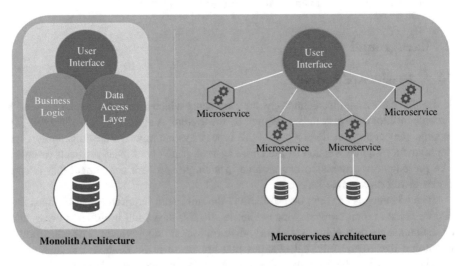

**Fig. 1.** Representation of the basic structure - Application with Monolithic Architecture and Micro services Architecture

## 2.3  KVM

The Kernel Virtual Machine – KVM, is an advanced virtualization technology that functions as a Linux process and is classified as a Hypervisor type I[1] [7].

Its administration is mainly done through the use of a Command Line Interface - CLI. Its allied application is QUEMU, also known as a generic and open source machine emulator and virtualize [8]. A KVM supports hot migration, which means that it allows physical servers or even Full Data Centers for maintenance without interrupting the Guest Operating System [9].

To facilitate the management of the KVM, the Virtual Machine Manager application brings a graphical interface to be used.

## 2.4  Containers

The containers are quite similar to virtual machines with the service these provide. The only exception is that they do not have an overhead that manages the execution of a separate kernel and the virtualization of all hardware components [9].

Containers enable each workload to have exclusive access to resources such as processor, memory, service account and libraries, which are essential for the development process [10]. In addition, they run as a group of isolated processes within an operating system. Like this, it optimizes startup times and easy maintenance. Each container is a package of libraries and dependencies necessary for its operation; therefore they area independent, hence the term of isolated processes.

One of the most critical risks in containers is the poor visibility of the processes that run inside a container with respect to the limit of resources which the host machine has [11].

As for safety, this is handled by management groups and namespaces permits. The users do not have the same treatment inside and outside the container.

Docker is broadly leading this technological segment as far as container handling is concerned. This Open Source technology has become the most commonly used tool by DevOps for container management in large-scale projects.

It has even been suggested as a solution for more interoperable cloud applications packaging [14, 15].

Docker initially using Linux Containers - LXC as default execution environment. However, changing distributions was a problem when trying to standardize a generic version. Docker, in version 0.9, includes its own runtime environment called LibContainer, which helped him become a standard and cross-platform technology Docker.

## 3  System Approach

This research proposes the implementation of an application with modern technology, in addition, to adapting environments where it can be executed and subjected to stress tests to calculate the level of performance.

---

[1] Works at the hardware level without an intermediary Operating System.

### 3.1    Application Design

The application uses modern programming technology based on Node.js, SQL Databases (MySQL) and NoSQL (MongoDB). The application represents a basic design - in Backend - for the management of forums chats, comments or notifications. It consists of user, thread and publication.

This application is based on a model published by AMAZON WEB SERVICES – AWS, in its public repository of GitHub with the name Node.js Micro services Deployed on Elastic Compute Cloud - EC2 Container Service.

The evolution of an App based on a Monolithic Architecture is represented, and the result is the same App using a Micro services Architecture. Additionally, the adaptation of virtual environments where the services will be raised and the tests will be executed is increased.

The first version of the application uses the monolithic architecture, the structure of the directory (project) and the arrangement of the files with the code, it is understood that all functionalities are grouped into a single resulting component (Fig. 2 shows the directory of the monolithic version).

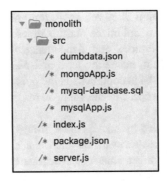

**Fig. 2.**  Structure of the Monolithic Application directory.

The second version is based on Micro services architecture. Both the structure of the directory and the distribution of the code in the files becomes different. Even giving place to its provision for its subsequent coupling with Docker containers (Fig. 3 shows the directory of the Micro services version).

### 3.2    Runtime Environment

Each version of the proposed application is executed in an environment that coincides with its trend. This is the case of the monolithic application that runs on a KVM and the application with the Micro services architecture opened on containers managed with Docker.

For both cases, hardware resources are of the same characteristics Intel Core i7 2.7 GHz CPU, 16 GB memory, HD disk and Virtual Network.

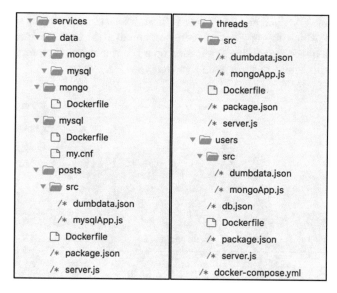

**Fig. 3.** Structure of the application directory with Micro services.

As per the Software, the base Operating System used for both, the host and the KVM, is Ubuntu 16.04, Long Term Support - LTS desktop and server respectively, QEMU 2.10, VMM 1.3.2, Docker 18.03.0-ce, Docker Compose 1.14, Java 1.8.0_162, NodeJS 6.14.1, MySQL 5.7.21, MongoDB 3.4.10, Apache JMeter 3.3, Server Agent 2.2 and NewRelic 2018.

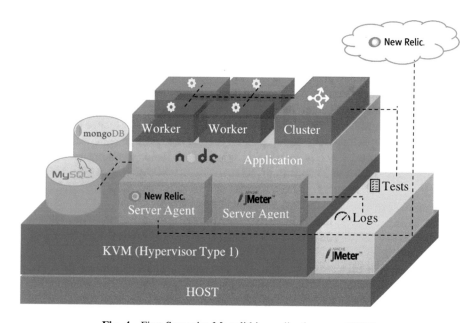

**Fig. 4.** First Scenario: Monolithic application on a KVM.

With JMeter, scripts are generated with stress tests, allowing it to specify the conditions to which the application will be submitted in each environment (Fig. 4 representation of the monolithic application on a KVM and Fig. 5 representation of the application with Micro services with containers).

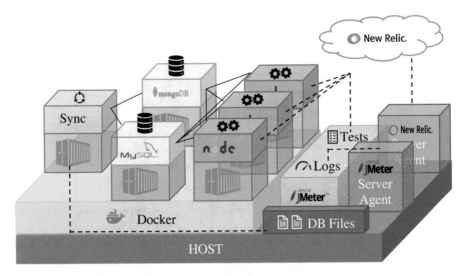

**Fig. 5.** Second Scenario: Application with Micro services on containers.

## 4    Assessments and Comparison

Stress tests are executed for each version of the application in its proper environment.

A series of tests are executed in a Script file generated with JMeter. It is configured to send asynchronous requests via HTTP to the application's Endpoints. Therefore, it simulates a number of connected users and defines a number of repetitions and threads. For the Monolithic Application, the KVM starts and then does the application. For the Application with Microservices, the containers are started using Docker that includes the start of the application. For stress tests, on one hand, there is the generation of data, and on the other hand the consumption (selection) of the same. The number of requests is limited to the processing capacity of the monolithic application because an increase in the number of requests generates errors of memory overflow or service drop (Application).

The tools (Software) JMeter, NewRelic and ServerAgent are used, which allow the collection of the "Logs", in files that are then processed and analyzed for their proper interpretation.

Two case of study are generated. The first with a reduced number of requests (273) to the application and the second case with a larger number of requests (1053). This number is a limit since increasing this amount generates memory errors and slows the response from the application.

## 4.1   CPU Consumption

Although, the performance of the application shows an advantage with the second scenario, it also evidences a greater consumption of the CPU resource. Indeed on average it consumes 67.73% versus 55.35% of the first scenario. Giving a difference of 12.38% in favor of the first scenario that generates less consumption of this resource.

## 4.2   Memory Consumption

Regarding this resource, it can be seen that on average there is a higher consumption. That is, 31.63% on the second scenario compared to 22.76% of the first scenario, with a difference of 8.87% in favor of this last scenario.

## 4.3   Network Bandwidth

In the case of the second scenario, the speed (kilobytes per second) is reduced compared to the first scenario. Because Docker generates a private network for the execution of its containers, the same one that serves as means of communication between them, thus reducing bandwidth capacity.

On average, the speed of the bandwidth in terms of receiving a KVM exceeds the containers by 85%. While the results improve in the speed of sending packages having a difference of 12.06%. For the purposes of the experiment, the average between these values is obtained, resulting in a 48.5% difference, where the first scenario exceeds the second scenario in terms of performance.

## 4.4   Disk Write and Read Speed

In both cases, the bandwidth is greater in the second scenario, which implies greater efficiency. For this case there is a difference of 78.5% in disk reading and 79.2% in disk writing respectively for each environment. Overall and for research purposes, it gives a final result of 78.85% in favor of the second scenario.

To get a better idea in another unit of measure, on average the first scenario makes disk write at a speed of 788.11 kilobytes per second and in the second stage the amount of data is 3.76 Megabytes per second.

## 4.5   Application Performance

In both cases and scenarios the requests end successfully, however, in the second scenario case 1 generates 2 server response errors (Tables 1 and 2).

**Table 1.** Timesheets duration of the test from the first to the last request

|        | Number of requests | Monolith time | Microservices time | Diff time | Diff % |
|--------|--------------------|---------------|--------------------|-----------|--------|
| Case 1 | 273                | 00:01:50      | 00:01:29           | 00:00:21  | 19.09% |
| Case 2 | 1053               | 00:14:17      | 00:12:08           | 00:02:09  | 15.05% |

**Table 2.** Table of number of requests processed per second

|        | Number of requests | Monolith | Microservices | Diff  | %      |
|--------|--------------------|----------|---------------|-------|--------|
| Case 1 | 273                | 2.5/s    | 3.1/s         | 0.6/s | 24%    |
| Case 2 | 1053               | 1.2/s    | 1.4/s         | 0.2/s | 16.67% |

On average, there is a reduction of 17.7% in total runtime using the scenario 2 with respect to containers Micro services and equivalent monolithic.

## 5    Conclusion and Future Work

The Monolithic Architecture, in conjunction with the KVM, show a favorable performance of CPU, RAM and Network. However, we must consider two new scenarios that a DevOps must take into account in the planning of a project. As per the Microservices, they prove to be efficient in the performance of resources, the Application and Disk. The two architecture are reasonably equal when the time is a relevant factor.

A shorter response time, means a higher consumption of resources, like in the specific case of Micro services with containers where usually large-scale computer systems, high flow and high performance information are driving. This scenario is convenient when the objective of the application is to process the greatest number of requirements in the shortest time possible, regardless of the limitation of resources and where scalability is a priority.

In longer response time there is a lower consumption of resources. This is a case specific to a monolithic application with KVM, where the computer system does not require a high level of resources and traffic is moderate or low.

Given the current trend, most computer systems look for ways to position themselves on top with the most modern technologies. The response time is almost imperceptible when accessing a technological service. The performance not only focuses on consuming a lower percentage of a resource, but also on optimizing so that consumption is adequate in shorter times.

Micro services and containers have been the subject of various investigations given their potential. This gives rise to new challenges, one of which is motivated to perform the same experiment proposed in this research but this time with services in the Cloud.

## References

1. Nielsen, C.D.: Investigate availability and maintainability within a microservice architecture (2015). http://cs.au.dk/fileadmin/site_files/cs/AA_pdf/ClausDNielsen_rapport.pdf
2. Sun, L., Li, Y., Memon, R.A.: An open IoT framework based on microservices architecture. China Commun. **14**, 154–162 (2017)
3. Kratzke, N.: About microservices, containers and their underestimated impact on network performance. In: ResearchGate, Lubeck, Germany (2015)
4. Fowler, M., Lewis, J.: Microservices. https://martinfowler.com/articles/microservices.html

5. Fussell, M.: Why a microservices approach to building applications? https://docs.microsoft. com/es-es/azure/service-fabric/service-fabric-overview-microservices

6. Newman, S.: Building Microservices: Designing Fine-Grained Systems. O'Reilly Media, Inc., Sebastopol (2015)

7. Felter, W., Ferreira, A., Rajamony, R., Rubio, J.: An updated performance comparison of virtual machines and linux containers. In: 2015 IEEE International Symposium on Performance Analysis of Systems and Software (ISPASS), pp. 171–172. IEEE (2015)

8. Bartholomew, D.: Qemu: a multihost, multitarget emulator. Linux J. **2006**, 3 (2006)

9. Prashant, D.: A survey of performance comparison between virtual machines and containers. 4, (2016)

10. Scott, J.: A practical guide to microservices and containers: mastering the cloud, data and digital transformation (2017)

11. Vaughan-Nichols, S.J.: New approach to virtualization is a lightweight. Computer **39**, 12–14 (2006)

12. Khazaei, H., Barna, C., Beigi-Mohammadi, N., Litoiu, M.: Efficiency analysis of provisioning microservices. In: 2016 IEEE International Conference on Cloud Computing Technology and Science (CloudCom), pp. 261–268 (2016)

13. Stubbs, J., Moreira, W., Dooley, R.: Distributed systems of microservices using docker and serfnode. In: 2015 7th International Workshop on Science Gateways, pp. 34–39 (2015)

14. Pahl, C.: Containerization and the PaaS Cloud. IEEE Cloud Comput. **2**, 24–31 (2015)

15. Gerlach, W., Tang, W., Keegan, K., Harrison, T., Wilke, A., Bischof, J., D'Souza, M., Devoid, S., Murphy-Olson, D., Desai, N., et al.: Skyport: container-based execution environment management for multi-cloud scientific workflows. In: Proceedings of the 5th International Workshop on Data-Intensive Computing in the Clouds, pp. 25–32. IEEE Press (2014)

# A Proposal for an Electronic Negotiation Platform for Tourism in Low-Density Regions: Characterizing a Functional Analysis and Prototype for the Douro Valley

Luís Cardoso[1], José Martins[1,2], Ramiro Gonçalves[1,2(✉)],
Frederico Branco[1,2(✉)], Fernando Moreira[3,4],
and Manuel Au-Yong-Oliveira[5]

[1] University of Trás-os-Montes e Alto Douro, 5000-801 Vila Real, Portugal
luis29-10-1995@hotmail.com,
{jmartins, ramiro, fbranco}@utad.pt
[2] INESC TEC (coordinated by INESC Porto), Faculty of Engineering,
University of Porto, 4200-465 Porto, Portugal
[3] Univ Portucalense, Portucalense Institute for Legal Research – IJP, Research
on Economics, Management and Information Technologies – REMIT,
Porto, Portugal
fmoreira@uportu.pt
[4] Universidade Aveiro, IEETA, Aveiro, Portugal
[5] GOVCOPP, Department of Economics, Management, Industrial Engineering
and Tourism, University of Aveiro, 3810-193 Aveiro, Portugal
mao@ua.pt

**Abstract.** Traditional e-Commerce platforms tend to mimic existing physical store processes in which customers are given the opportunity to purchase only the available products without any type of personal customization or even without the possibility of being allowed to present a desired product/service to which the market might be able to respond. With this in mind, a prototype for an electronic negotiation platform directed at the tourism sector has been developed and focused on Douro Valley tourism operators, products and services. In our opinion this prototype might help to develop new business models drawn from the customers' willingness to have a personal tourism experience and not just something off the shelf.

**Keywords:** e-Commerce platform · Tourism · Negotiation platform
e-Commerce prototype · Software analysis and prototyping

## 1 Introduction

The impact of information and communication technologies (ICT) on companies in the various branches of industry has never been greater. In fact, finding a product or service that has not been altered or influenced by ICT in any way is becoming more and more difficult [1].

© Springer Nature Switzerland AG 2019
J. Mejia et al. (Eds.): CIMPS 2018, AISC 865, pp. 280–292, 2019.
https://doi.org/10.1007/978-3-030-01171-0_26

Since the emergence of the Internet, the way business is being conducted has been changing. ICT has opened up new opportunities and threats that have changed traditional ways of doing business. Consequently, the processes involved in various fields such as trade, economy, etc., have also changed [2]. ICTs, in constant evolution, have thus become one of the decisive factors for a good implementation and subsequent competitive positioning of organizations. In this sense, the Internet has affirmed itself as an important channel of communication and commercialization of goods and services.

Since any company can afford the costs associated with its insertion in the Internet/E-Commerce (EC), it has become possible to compete in this global economy, and consequently expose and sell products/services to a wider range of customers.

Without question, with the increasing adoption of EC in the world and its exceptional potential, it is now clearly evident that countries where ICT are more developed have an advantage. EC is beginning to be considered as a new form of business, with extensive scope to create new strategies and business opportunities, either at the level of organizations (it allows for the reduction of costs and process improvements), or at the level of the communication channel with the customer (it allows new sales channels, new products and services, new forms of relationship and new opportunities in new business projects) [3].

In fact, the rapid development of the Internet has created a great opportunity to conduct business activities electronically. Customers can search for information about a product on the web and shop online. However, the price or terms for the goods or services are usually pre-specified by the seller or determined by well-defined procedures such as online auctions.

Although the traditional process of buying and selling on Web platforms is already quite stable, the latest changes in the way consumers approach online searches and purchases of products and services has forced organizations, and by default, EC platforms to change their ways of selling.

As e-commerce begins to become more sophisticated, it becomes necessary for more complex support contracts or business options to be mutually determined (such as negotiating the terms of a tourist trip, including price, date of arrival, date of departure, etc.).

In a world where more and more businesses are being moved to the Internet, disputes also increase in the virtual space, so the demand for trading also stems from the need to resolve a dispute. One solution would be to adjust e-commerce platforms to allow customers not only to buy products/services, but also to negotiate their terms in order to customize them according to their preferences.

In this context, the objective of this research is to achieve a Marketplace e-commerce Web platform, where the customer can propose the products or services he/she would like to have and the suppliers registered for such requests can respond to those proposals. This new form of negotiation and online bidding, in addition to the traditional way of selling, will transform EC platforms into tools that are much better suited to new customers.

## 2  Conceptual Background

### 2.1  Electronic Commerce and Negotiation Platforms

ICTs, in constant evolution, have become an indispensable tool for conducting business and for achieving effective competitive positions by organizations.

In fact, since the emergence of the Internet, the way to conduct business has been changing quite significantly. Due to the characteristic of ubiquity, associated with the Internet, the market is able to be free of restrictions to physical space as it is possible to make purchases from home, at work, or from any other site, using the technologies of mobile commerce. The result of this feature is a commercial space, that is, a market space that extends beyond traditional frontiers and away from a temporal and geographical location [2].

Based on what has been described, from a consumer's point of view, ubiquity reduces the costs of a transaction (costs of market share). To make transactions, it is no longer necessary to invest time and money on trips to a given market. From the point of view of the negotiator, since almost any company can afford the costs associated with its insertion in the Internet, it becomes possible to compete in this global economy, and consequently expose a firm's products/services to a wider range of clients.

As the global market expands and business and personal relationships are increasingly taking place in an online space, it is also becoming more commonplace and 'normal' to conduct negotiation processes electronically [4].

With the evolution of ICT and the emergence of the Internet, new opportunities for the design and implementation of software capable of supporting negotiators, mediators and arbitrators, have appeared and so currently electronic trading platforms are widely used [5].

Before describing electronic trading systems, it is considered important to first define the trading concept.

A negotiation can be understood as a process of communication between a group of parties, with a conflict of interests or preferences, that aims at reaching an agreement or compromise between the parties [6, 7]. Negotiation processes occur in a myriad of ways, being influenced by ethical, cultural, and social circumstances [8].

Electronic Trading refers to negotiation processes that are fully or partially conducted through the use of electronic devices, which use digital channels to carry data [9].

An electronic trading system is a system that uses the Internet to facilitate, organize, support and/or automate trading activities [10, 11]. This definition also includes all types of software capable of helping one or more negotiators or mediators, which includes email, chat [12, 13], software that combines negotiation and bidding mechanisms [14], among others. Typically, these systems adopt a Web-based design and are deployed on the Internet [15].

According to Braun et al. [8], an electronic trading system must have at least one or more of the following capabilities:

- To support the concession and decision making;
- To support the suggestion of offers and the negotiation of agreements;

- To support access and the critique of offers while also allowing counter-offers;
- To support the structuring and organization of processes;
- To provide information and knowledge;
- To facilitate and organize communication;
- To assist in the preparation of an agreement;
- To provide access to specialists, mediators or facilitators.

## 2.2    Electronic Commerce and Negotiation Platforms Applied to Tourism

The adoption of ICTs in the tourism sector over the last few years, which represents a strong technological evolution, also motivated a different approach by tourists, who are increasingly sophisticated and who expect more dynamic tourism experiences [13]. It is therefore crucial that tourism organizations adapt their business in a way that keeps them competitive [14].

Contemporary society has made the tourism sector highly information-intensive, as ICT increasingly plays a determining role in the success of industry organizations. The use of ICT in the tourism sector not only created new consumption habits but also created new opportunities for agents.

As the existing literature shows, e-commerce is a very important tool not only for the development of organizations, but particularly for the development of territories, and this relevance is especially important in regions where a significant set of constraints already exist, as is the case in low density regions [2].

Over the last few years, several authors have focused on the formalization and conceptualization of the concept of electronic commerce, but mainly on the possibilities associated with its application in contexts where economic, geographic or social factors may represent, in some way, an obstacle to the so-called traditional marketing of goods and services [7].

According to Cao et al. [16], the incorporation of complementary negotiation actions into the typical process of buying and selling through digital mechanisms has been the target not only of the attention of the scientific community [9, 10], but also of organizations that see an opportunity to be able to buy products/services at a more cost-effective price, and to be able to offer their customers the ability to negotiate/customize the product or service they are trying to acquire, and in real time get feedback on their proposals [12].

## 2.3    e-Commerce and Low-Density Regional Tourism

Such as Boateng et al. [17] claimed a decade ago, and Awiagah et al. [18] currently confirm, e-commerce when used correctly can serve as a catalyst for the economic and social development of a region, as it allows for the range of economic operators to extend themselves virtually and at much lower costs than associated with traditional export initiatives.

The incorporation of e-commerce into (more or less developed) business activities related to tourism is assumed to be the key element for the sector's change and the assumption of the importance that the international dimension has for organizations and regions. This process of digitization, through the incorporation of mechanisms for the promotion and online sale of products and goods, is seen as more relevant and decisive in regions with lower levels of economic and social development [19].

Over the last few years, tourism has been evolving, proving the existence of new sectoral niches where the elements "inequality" and "disadvantage" are the support for a new set of tourism offers widely stimulated and mediated through digital platforms and whose sale is typically conducted online [20].

According to Chen and Tsao [21] and Rodrigues et al. [22], the use of electronic commerce as a mechanism to stimulate tourism is even more critical when in rural environments, as is the case in the Douro region, since the typical lack of capacity (technical, technological, functional, financial, etc.), greatly limits the business actions of tourism organizations located in these environments and consequently their ability to attract new customers and reach new markets.

## 3   Functional Analysis of an Electronic Negotiation Platform for Tourism

The identification and characterization of requirements refers to the process of verifying a given system's necessary features and its users' needs. Considering Anu et al. [23] argument concerning requirements analysis, a decision was made to identify, in a collaborative manner, all requirements inherent to an electronic negotiation platform that could serve the tourism sector, thus assuming that the developed system will achieve the necessary quality.

In order to formalize the software engineering process, we have performed a series of efforts towards analyzing the system to be developed. These efforts have resulted in the definition of the system architecture, a set of use cases (generalized into one single use case diagram representing a global perspective on the system's features and interactions) and a class diagram (represented in this manuscript by a simplified version that conceived the main entities supporting the system).

### 3.1   Architecture Proposal

A straightforward analysis of existing e-commerce and electronic negotiation platforms allowed us to perceive a common architecture alignment between all of them. Therefore, by implementing a collaborative system architecture design where the main Douro Valley stakeholders have been uniting efforts and giving their own inputs on what an electronic negotiation platform for tourism (in the referred region) should be, an architecture proposal has been reached (Fig. 1).

Considering that the work behind the present paper is an ongoing research effort, a previous version of the negotiation platform has already been subject to peer-review and has been considered extremely interesting for serving as the basis for the development of a software piece to mimic the presented component connections.

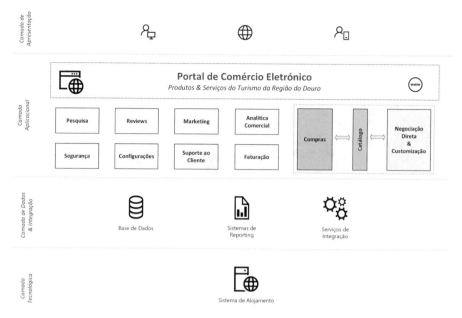

**Fig. 1.** Electronic negotiation platform architecture. Adapted from [22].

## 3.2   Global Perspective on the System's Business Use Cases

According to Huanca and Oré [24], the creation of business use cases that mimic the real use of the system to be developed might be considered one of the most important tasks associated with the analysis and specification of software (aka "software engineering"). Following these indications, we have identified a set of use cases that represent the main features that the electronic negotiation platform will involve and that are going to be made available on the platform we are going to implement.

As one can perceive in Fig. 2, we have identified four main user profile interactions with the system: the system administrator (with access to all available features), the operators (representing the tourism operators that are going to sell their products or services on the platform), the customers (who are those not only purchasing the available products but also presenting proposals for changes to existing products or to new products), and the visitor (who will in fact visit the region).

## 3.3   Global Perspective on the System Class Diagram

In order to understand the relations between the data objects that are going to be the basis of the electronic negotiation platform that we are going to develop, we have performed an analysis of the identified functional requirements and business use cases and were able to reach a set of conceptual classes (Fig. 3) and inherent relations.

By analyzing Fig. 3 we can perceive that for the platform we aimed to develop the main focus is the available offers (tourism products or services) and possible changes to

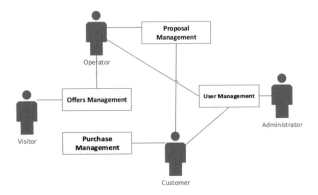

**Fig. 2.** Global perspective of the main business use cases identified.

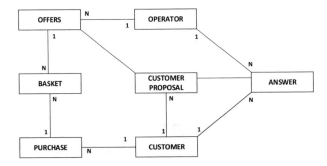

**Fig. 3.** Global approach to the electronic negotiation platform class diagram.

these offers that customers are willing to suggest, or even new customer proposed products to which registered operators can answer, thus the negotiation process.

## 4 Proposal of an Electronic Negotiation Platform for Tourism

### 4.1 Technological Solution Used

An online store, or a website that allocates an online store, can be implemented in one of two ways: (1) using content management system (CMS) systems; or (2) developing a customized solution from scratch.

If the e-store was built using CMS, the building process would be more linear and faster. However, given that the platform to be proposed has to implement a mechanism to support electronic negotiation between clients and vendors, and since in the light of our knowledge there are no modules for CMS that support this type of negotiation, we opted to develop an electronic trading platform for tourism (thus, a customized solution from scratch, as mentioned above).

In this sense, and after a systematic analysis of the various tools and approaches available for the development of responsive web platforms, ASP.NET MVC (Model View Controller) was chosen as the base programming framework and the Microsoft Visual Studio tool as the development support instrument. ASP.NET is Microsoft's platform for developing Web applications in .NET, which allows you to generate pages that contain HTML, CSS, and JavaScript that contain server-side functionality. The relational database supporting the entire Web platform developed was implemented in a Data Base Management System (DBMS).

As can be seen in Fig. 4, from the home page, a client can access the various zones of the developed platform. These areas are coloured in blue or red. Areas with a blue colour represent the pages that a client can access without being authenticated by the system. The coloured areas in red refer to the pages where the client needs to authenticate to access their content. For a client to log in to the system, he or she can do one of the following. If he or she already has an account, it is necessary to enter account information on the login page. This data can be queried and changed when querying the profile after the authentication in the system has been made. If there is no account, it is necessary to access the registration page and create a new account.

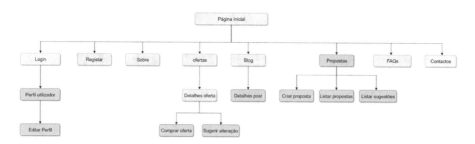

**Fig. 4.** Proposed prototype front-office navigation scheme.

The About page is intended to provide the customer with some information regarding the Douro area. In the zone of offers it is possible to consult all the active offers existing in the system for the various sectors of tourism. After seeing the details of the offer, the client can buy the tourist offer, or in case something does not suit his or her preferences, he or she can try to negotiate the same offer, suggesting some adjustments to it. All of these suggestions for change will be listed in the Proposals area. In addition, in this area, the client can also create and consult his or her own tourist proposal, which will be seen by all the operators belonging to the same sector. From the homepage, a client can also access the blog, where posts can be consulted which were created by the system administrator with news, curiosities, etc. The FAQs page is intended to help the customer with some questions regarding the platform by presenting some of the frequently asked questions. If the customer wants to contact the entity responsible, he or she can find all of the necessary information in the Contacts area.

Figure 5 represents the home page of the developed platform. This page has as its main objective to present the platform to the client or potential clients. Since this will be the main navigation page of the platform, as well as the first page that potential customers will come into contact with, we chose to build it with a simple yet attractive design that works on any electronic device. It was also considered important to make the homepage dynamic (so that the information presented is always up to date), interactive and easy to navigate. An important part for browsing within the website is the menu. This menu consists of some important sections on the website such as information about the Douro region, tourist offers, and the blog, among others. In order to improve the user experience on the platform, it was decided to place several primary and secondary call-to-actions, such as "View Prices", "View Details", among others, in order to deepen and direct the users' navigation.

**Fig. 5.** Proposed prototype homepage layout.

Figure 6 represents the listing of tourist offers. Here a customer can search and consult all the existing tourist offers in the system. It was decided to split this page into two sections. In the first section, located on the left side, it was decided to provide a search panel, which allows the client to search the different tourist offers according to his or her criteria. The number of offers found is shown above. If the customer cannot find the tourist offer he or she is looking for, it is possible to create a proposal by clicking on the "Create Proposal" button. In the second section, located on the right hand side, is the list of tourist offers found. The information on the tourist offers listed is summarized, where only the name of the proposal is presented, as is a brief description of it, along with the price and the number of comments it has. If the client has an interest in any of the tourist offers, he or she simply has to click on the "Details" button to see more information about the offer and, if they exist, the comments on the offer.

Figure 7 is presented when the client wants to consult the details of a tourist offer. Here the customer can consult more information about the offer, such as more images, comments made by other customers, as well as contact information of the operator responsible for the tourism offer in question. The customer can also give his or her opinion when commenting on the tourist offer or when responding to another customer's comment. In case the customer is interested in the tourist offer he or she can proceed with the purchase by clicking on the "buy" button. If there is something that does not suit the possible future customer, there is an option to negotiate with the operator responsible for this offer. To do this, the client must press the "Suggest Change" button.

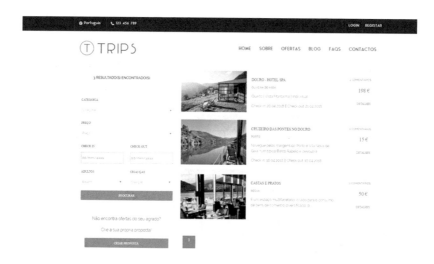

**Fig. 6.** Proposed prototype product listing page layout.

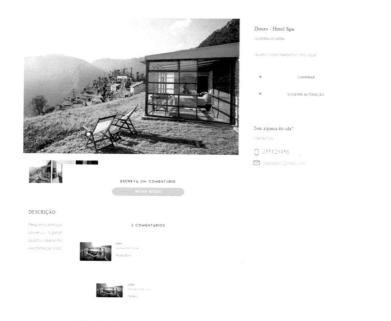

**Fig. 7.** Proposed prototype product edit page.

When the customer clicks on the "Suggest Change" button, a form is shown (Fig. 8). When this form is opened, it will automatically be filled-in with the details of the tourist offer in question. Here the customer can consult the details of the offer and change some of the fields such as the price, the number of adults and children, among others. Note that some fields cannot be changed such as the title or description.

**Fig. 8.** Proposed prototype product change proposal.

When the customer considers that he or she has made the necessary changes to make the offer fit his/her needs, then one simply has to click on the "Submit" button to send the change suggestion. If he/she clicks on the "Cancel" button, the changed values are no longer valid, and the change suggestion is not sent. Once submitted, this suggestion for change can be consulted by the operator of the offer in question.

## 5   Conclusions and Future Work

The developed prototype intends to be an initial step towards the validation of the bi-directional e-Commerce paradigm applied to tourism, and particularly to tourism in low density regions. By drawing its developed features from a previously performed funcional and technical analysis, where software engineering techniques have been used, we believe that the presented output represents a step forward towards empowering even more the tourism operators of regions such as the Douro Valley, where despite the low levels of digital capabilities there is a proven interest in applying new technologies to the tourism sector as a way to address the new tourists' necessities.

In the future we would like to validate the developed prototype with a controlled test where real Douro Valley tourism operators are incorporated and pre-controlled samples of tourists are invited to be a part of the trial. This will allow us to perceive, first-off the acceptance level that both operators and tourists have for these technologies, and secondly the alignment between the developed solution and the actors' real necessities.

**Acknowledgements.** This paper was developed under Project I&D DOUROTUR – Tourism and technological innovation in the Douro, operation number NORTE-01-0145-FEDER-000014, co-funded by Fundo Europeu de Desenvolvimento Regional (FEDER) through NORTE 2020 (Programa Operacional Regional do Norte 2014/2020).

# References

1. Demirkan, H.: Special section: enhancing e-commerce outcomes with IT service innovations. Int. J. Electron. Commer. **19**(3), 2–6 (2015)
2. Laudon, K.C., Traver, C.G.: E-commerce. Pearson (2013)
3. Hortala-Vallve, R., Llorente-Saguer, A., Nagel, R.: The role of information in different bargaining protocols. Exp. Econ. **16**, 88–113 (2013)
4. Brennan, L.L.: Computer-Mediated Relationships and Trust: Managerial and Organizational Effects. IGI Global, Hershey (2007)
5. Kersten, G.E., Lai, H.: Negotiation support and e-negotiation systems: An overview. Group Decis. Negot. **16**, 553–586 (2007)
6. Luo, X., Jennings, N.R., Shadbolt, N., Leung, H.-F., Lee, J.H.-M.: A fuzzy constraint based model for bilateral multi-issue negotiations in semi-competitive environments. Artif. Intell. J. **148**, 53–102 (2003)
7. Luo, X., Sim, K.M., He, M.: A knowledge based system of principled negotiation for complex business contract. In: International Conference on Knowledge Science, Engineering and Management, pp. 263–279. Springer, Heidelberg (2013)
8. Braun, P., Brzostowski, J., Kersten, G., Kim, J.B., Kowalczyk, R., Strecker, S., Vahidov, R.: E-negotiation systems and software agents: methods, models, and applications. In: Intelligent Decision-Making Support Systems, pp. 271–300. Springer, Heidelberg (2006)
9. Kersten, G.: The science and engineering of e-negotiation: an introduction. In: Proceedings of the 36th Annual Hawaii International Conference on System Sciences, 2003, pp. 27–36. IEEE (2003)
10. Bichler, M., Kersten, G., Strecker, S.: Towards a structured design of electronic negotiations. Group Decis. Negot. **12**, 311–335 (2003)
11. Insua, D.R., Holgado, J., Moreno, R.: Multicriteria e-negotiation systems for e-democracy. J. Multi-Criteria Decis. Anal. **12**, 213–218 (2003)
12. Lempereur, A.P.: Innovation in teaching negotiation towards a relevant use of multimedia tools. Int. Negot. **9**, 141–160 (2004)
13. Moore, D.A., Kurtzberg, T.R., Thompson, L.L., Morris, M.W.: Long and short routes to success in electronically mediated negotiations: group affiliations and good vibrations. Organ. Behav. Hum. Decis. Process. **77**, 22–43 (1999)

14. Teich, J.E., Wallenius, H., Wallenius, J., Zaitsev, A.: Designing electronic auctions: an internet-based hybrid procedure combining aspects of negotiations and auctions. Electron. Commer. Res. **1**, 301–314 (2001)
15. Kersten, G., Lai, H.: Electronic negotiations: foundations, systems, and processes. In: Handbook of Group Decision and Negotiation, pp. 361–392. Springer, London (2010)
16. Cao, M., Luo, X., Luo, X.R., Dai, X.: Automated negotiation for e-commerce decision making: a goal deliberated agent architecture for multi-strategy selection. Decis. Support Syst. **73**, 1–14 (2015)
17. Boateng, R., Heeks, R., Molla, A., Hinson, R.: E-commerce and socio-economic development: conceptualizing the link. Internet Res. **18**, 562–594 (2008)
18. Awiagah, R., Kang, J., Lim, J.I.: Factors affecting e-commerce adoption among SMEs in Ghana. Inf. Dev. **32**, 815–836 (2016)
19. Bui, T., Yen, J., Hu, J., Sankaran, S.: A multi-attribute negotiation support system with market signaling for electronic markets. Group Decis. Negot. **10**, 515–537 (2001)
20. Rangaswamy, A., Shell, G.R.: Using computers to realize joint gains in negotiations: toward an "electronic bargaining table". Manag. Sci. **43**, 1147–1163 (1997)
21. Chen, Y., Tsao, H.: A comparison of approaches of poverty alleviation through e-commerce. In: 2017 3rd International Conference on Information Management (ICIM), pp. 78–82. IEEE (2017)
22. Rodrigues, S., Goncalves, R., Teixeira, M.S., Martins, J., Branco, F.: Bidirectional e-commerce platform for tourism in low-density regions: the Douro Valley case study. In: 2018 13th Iberian Conference on Information Systems and Technologies (CISTI), pp. 1–5 (2018)
23. Anu, V., Hu, W., Carver, J., Walia, G., Bradshaw, G.: Development of a human error taxonomy for software requirements: a systematic literature review. Inf. Softw. Technol. (2018, in press)
24. Huanca, L., Oré, S.: Factores que afectan la precision de la estimacion del esfuerzo en proyectos de software usando Puntos de Caso de Uso. RISTI-Revista Ibérica de Sistemas e Tecnologias de Informação, 18–32 (2017)

# Correction to: Framework for the Analysis of Smart Cities Models

Elsa Estrada, Rocio Maciel, Adriana Peña Pérez Negrón,
Graciela Lara López, Víctor Larios, and Alberto Ochoa

**Correction to:**
**Chapter "Framework for the Analysis of Smart Cities Models"**
**in: J. Mejia et al. (Eds.):** *Trends and Applications in Software*
*Engineering*, **AISC 865,**
**https://doi.org/10.1007/978-3-030-01171-0_24**

In the original version of the book, the following belated correction have been incorporated: The author name "Graciela López Lara" has been changed to "Graciela Lara López" in the Frontmatter and in Chapter "Framework for the Analysis of Smart Cities Models".

---

The updated version of this chapter can be found at
https://doi.org/10.1007/978-3-030-01171-0_24

© Springer Nature Switzerland AG 2019
J. Mejia et al. (Eds.): CIMPS 2018, AISC 865, p. E1, 2019.
https://doi.org/10.1007/978-3-030-01171-0_27

# Author Index

© Springer Nature Switzerland AG 2019
J. Mejia et al. (Eds.): CIMPS 2018, AISC 865, pp. 293–294, 2019.
https://doi.org/10.1007/978-3-030-01171-0

Printed in the United States
By Bookmasters